Technologies for Personalised Healthcare

This book explores how personalised healthcare can be realised with the integration of new and established technologies, and how novel methods can be developed to improve health outcomes in lifestyle-related conditions.

Technologies for Personalised Healthcare: Enabling Patient-Centred Approaches in the Management and Prevention of Lifestyle-Related Conditions explores various aspects of personalised healthcare in the context of lifestyle-related conditions—from how statistics and machine learning can be used to extrapolate clinically useful stratifications from big data, to how adaptive AI can enable behaviour change; from how biosensors and point-of-care devices can bring diagnostics closer to the patient, to the ethical aspects of personalised medicine. The highly multidisciplinary nature of the book gives the reader a broad view of established and emerging technologies for personalised healthcare. It provides the technical background underpinning these technologies, while case studies present practical applications of the methods discussed. The content and worked examples found within this book can serve as a reference guide on the technologies and methods needed to tap into the endless opportunities to be realised within personalised healthcare.

This book is intended for researchers and academics who want to know more about how they can apply computational methods and electronics within the field of personalised healthcare. It is also for those who are interested in the bridge between technology, engineering, and medicine.

It is our hope that this book will serve as a catalyst for further innovation and collaboration in the pursuit of personalised healthcare solutions.

Technologies for Personalised Healthcare

Enabling Patient-Centred Approaches in the Management and Prevention of Lifestyle-Related Conditions

Francesca Romana Cavallo and
Christofer Toumazou

CRC Press
Taylor & Francis Group
Boca Raton London New York

CRC Press is an imprint of the
Taylor & Francis Group, an **informa** business

Designed cover image: © Shutterstock

First edition published 2025
by CRC Press
6000 Broken Sound Parkway NW, Suite 300, Boca Raton, FL 33487-2742

and by CRC Press
4 Park Square, Milton Park, Abingdon, Oxon, OX14 4RN

CRC Press is an imprint of Taylor & Francis Group, LLC

ISBN: 978-1-032-73607-5 (hbk)
ISBN: 978-1-032-73593-1 (pbk)
ISBN: 978-1-003-46498-3 (ebk)

DOI: 10.1201/9781003464983

Typeset in Times
by Deanta Global Publishing Services, Chennai, India

Contents

About the authors

Francesca Romana Cavallo earned her PhD from Imperial College London, focusing on technologies for personalised healthcare. She also holds an MEng from the same institution. Currently, she works in the healthcare field, leveraging her expertise to drive innovative solutions and advance the field of healthcare technology.

Christofer Toumazou is Regius Professor of Engineering at Imperial College London. He is a Fellow of the Royal Society, the Royal Academy of Engineering, and the Academy of Medical Sciences. In recognition of several innovations, he was awarded the 2014 European Inventor Award, the Royal Society Gabor Medal, and the IET Faraday Medal.

Preface

Writing this book has been simultaneously a great challenge and a great source of satisfaction. It encapsulates six years of research and experimentation in the field of technologies for personalised healthcare. Personalised medicine, which tailors treatments and interventions to the individual, involves the integration of multiple technological advancements. This book explores various technologies that can enable personalised healthcare, from how statistics and machine learning can be used to extrapolate clinically useful stratifications using big data, to how adaptive AI can enable behaviour change, to how biosensors and point-of-care devices can bring diagnostics closer to the patient. The many ethical and regulatory considerations involved in personalised medicine and healthcare are also discussed. We hope that the content and worked examples found within this book can serve as a reference guide to the technologies and methods needed to exploit the endless opportunities to be realised within personalised healthcare. We hope that clinicians, scientists, engineers, and interested lay readers will find this book useful.

1 The rise of personalised healthcare

INTRODUCTION

We are currently living through the Fourth Industrial Revolution, where new technologies are developed at unprecedented speed and data are generated at an increasingly faster pace. The integration of technologies such as artificial intelligence (AI), high-throughput sequencing, and smart sensors has blurred the lines between the physical, digital, and biological domains. As these technologies become more accurate and accessible, they are set to change how humans manage their lives and health.

Through these innovations, we are now able to quantify clinical, biometric, and behavioural data from daily life, and to analyse such data to yield impactful insights into human health. The analysis of large medical and biological datasets constantly uncovers associations between environment and disease, and elucidates the mechanisms and pathways underlying common and rare conditions. Further applications of big data analytics in healthcare include the prediction of clinical outcomes and disease progression, the identification of risk groups, drug discovery, and last but not least, personalised healthcare.

Integrating all of the data collected both in clinical settings and in daily life can greatly help in the fight against diseases caused by poor lifestyle practices. Non-communicable diseases (NCDs), which include diabetes and cardiovascular disease (CVD), are a significant burden to the modern world, accounting for more than 70% of deaths worldwide.[1] The highest proportion of all-cause disability-adjusted years is due to behavioural risks, such as low physical activity and poor diet, metabolic risks including increased blood pressure and high blood glucose levels, and the interaction between metabolic and behavioural risks.[2] Due to the very nature of these conditions, lifestyle medicine is the first-line therapy to manage the disease and prevent the onset of comorbidities.

Lifestyle medicine aims to prevent or manage diseases by targeting modifiable risk factors associated with lifestyle using evidence-based interventions to support beneficial changes in health behaviours.[3] However, despite the well-documented benefits of increasing physical activity, decreasing sedentary time, and reducing the consumption of certain foods, trial results of lifestyle interventions are challenging to replicate in the real world, mainly due to low participation rates.[4] Even though lifestyle medicine can clearly benefit from the integration of new technologies and despite the constant innovation, data availability, and computational capacity, one of the biggest challenges is the implementation of healthcare discoveries and innovations in real life. The acceptance and adoption of new technologies are not

DOI: 10.1201/9781003464983-1

straightforward, especially in the healthcare sector—it takes, on average, 17 years for research discoveries to be introduced in clinical practice.[5] For example, despite the accuracy and performance of algorithms, AI is currently not widely used in clinical settings. One reason is the *black-box* nature of many algorithms, which hinders patients' and practitioners' trust due to the inability to interpret the results. Another example is the lack of use of genomic sequencing and genotyping in clinical practice, despite the huge level of innovation that, in recent years, has rendered these technologies affordable and easily available. While advanced genomic techniques are increasingly available at the point of care, barriers such as the lack of evidence on the clinical usefulness of genomic testing prevent its wide adoption in healthcare.[6]

The real-life acceptability of technology becomes especially significant in the context of NCDs related to lifestyle, in which the prevention and management are, to an extent, in the hands of the individual. The prevention and management of these diseases require not only new biological discoveries but also accessible and user-friendly technologies that can enable the delivery of therapies and increase patients' self-efficacy and adherence to treatments. The effectiveness and adoption of lifestyle medicine can be greatly enhanced by new technologies able to provide personalised therapies and delivery at the point of care, improving motivation, adherence, and, therefore, health outcomes.

This book explores various aspects of personalised lifestyle and healthcare—from how statistics and machine learning can be used to extrapolate clinically useful stratifications using big data, to how adaptive AI can enable behaviour change; from how biosensors and point-of-care devices can bring diagnostics closer to the patient, to the ethical and regulatory aspects of personalised medicine. The content and worked examples found within this book can serve as a reference guide on the technologies and methods needed to tap into the endless opportunities to be realised within personalised healthcare.

Nowadays, the availability of data and technology can enable the realisation of solutions that are integrated with everyday life and can make a lasting positive impact on health. This can be achieved through the integration of multiple data sources, including clinical and -omics data, as well as data from wearable sensors and smartphones. As healthcare moves towards personalised and precision medicine, realising a truly integrated and holistic version of personalised healthcare is an exciting multidisciplinary endeavour with never-ending possibilities.

PERSONALISED MEDICINE AND PERSONALISED HEALTHCARE

Modern medicine is built on evidence derived from population studies. Observational studies may try to infer the effect of environmental factors on certain diseases by analysing samples that are representative of a population, and conclude that, on average, smoking is related to cancer. Another example is randomised controlled trials (RCTs), such as the one conducted by Lewis et al.,[7] that were conducted to evaluate the efficacy of sertraline and prove that, on average, it is effective in the treatment of depression. Population-level studies proved the negative association between smoking and cancer, and the effectiveness of sertraline for depression for the average population. But why doesn't every smoker develop cancer, and why do some people develop side effects after taking sertraline? There is indeed great variation in the individual response to treatment, or in the effect that a risk factor has for different people. A simple but significant example that illustrates the shortcomings of population—rather than individualised—approaches is the sex disparity in adverse drug events. Women are twice as likely to experience adverse drug events,[8] as most of the currently available drugs have been tested with RCTs performed on men and, therefore, women are prescribed dosages that have been calibrated on a male population. Without even considering individual differences relating to genetics or behaviour, we can see how individual factors (in this case sex) can influence drug response. In this simple example of drug adverse events caused by sex-related differences, adjusting drug dosages by taking into account the patient's sex could significantly improve patient outcomes.

Personalised medicine aims to overcome the limitations of one-size-fits-all approaches by tailoring preventive and therapeutic strategies to the individual, as depicted in Figure 1.1. European Union health ministers defined personalised medicine as

> a medical model using characterization of individuals' phenotypes and genotypes (e.g. molecular profiling, medical imaging, lifestyle data) for tailoring the right therapeutic strategy for the right person at the right time, and/or to determine the predisposition to disease and/or to deliver timely and targeted prevention.[9]

This definition covers not only the idea of personalised treatments but also the concept that disease risk is affected by the individual's genetic makeup and the environment they are immersed in. The definition also touches on delivering timely and targeted prevention, highlighting that proactive, as well as reactive, measures can be tailored to the individual and their associated risk of disease.

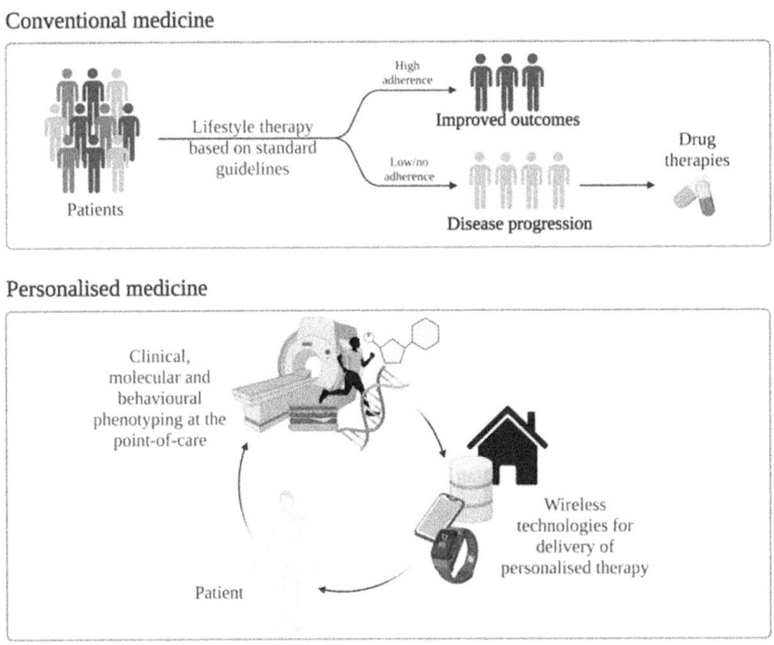

FIGURE 1.1 Conventional medicine and personalised medicine. Created in BioRender. Cavallo, F. (2024) BioRender.com/w35178.

Personalised medicine can also be considered as part of "P4 medicine": predictive, personalised, preventive, and participatory. First described by Hood et al.,[10,11] P4 medicine includes the concepts of prediction, personalisation, and prevention commonly associated with personalised medicine. However, it adds an element that is often forgotten when discussing the topic: participation. A core idea of personalised medicine is that patients increasingly take control of their own health.[12] But to do so, patients need to be educated on P4 medicine, its benefits as well as its risks. It is key that patients are well informed about the value of their data and the benefits of sharing them to advance biomedical research. This should come hand-in-hand with security, privacy, policy, and regulatory aspects that need to be firmly in place so as to not undermine patient trust and void any educative efforts to encourage patients to share their personal information. Furthermore, participation plays a primary role in closing the personalised medicine loop, as patient and consumer feedback allows the development of effective products and services that deliver personalised healthcare.

P4 medicine is not the only synonym of personalised medicine, and terms like "precision medicine," "individualised medicine," "stratified medicine," and "personalised healthcare" are often used interchangeably, leading to ambiguity. Patrinos et al.[13] asked ChatGPT, an AI-based chatbot based on a conversational language model developed by OpenAI, "What is the difference between precision medicine and personalized medicine?" ChatGPT replied with,

> Precision medicine and personalized medicine are often used interchangeably, but there is a subtle difference. Precision medicine refers to the use of genetic and other molecular information to tailor treatment plans to specific subgroups of patients, while personalized medicine takes into account a broader range of factors, such as lifestyle and environmental factors, in addition to genomic information, to tailor treatment plans to each individual.

Personalised medicine is a vast discipline, and despite efforts to pinpoint a definition, it can remain ambiguous. One of the reasons for such equivocality is the tendency to identify precision medicine with pharmacogenomics and genomics medicine. Pharmacogenomics is defined as the study of the inherited basis of the interindividual differences in the effects produced by the administration of a drug.[14] In simpler terms, it is the study of how genetic variations affect a patient's response to a particular medication.[15] For the US National Cancer Institute, precision medicine is a "form of medicine that uses information about a person's own genes or proteins to prevent, diagnose, or treat disease,"[16] pinpointing genetics as a core aspect of personalised medicine. Personalised medicine is not a new concept, but the latest wave of interest for personalised medicine was fuelled by big human genome sequencing studies like the Human Genome Project,[17] the Phase 1 HapMap project,[18] and the ENCODE project[19] in the early 2000s. Since then, decades of research have identified—and continue to uncover—genetic variations that influence disease risk, disease progression, or treatment response. However, genetics is not the only factor shaping health. Lifestyle-related conditions like type 2 diabetes (T2D), CVD and some types of cancers demonstrate that the environment, as well as genetic predisposition, affects the chances of developing the condition and its prognosis. The combination of social, economic, and physical environments, such as income and social status, education, personal behaviour, access to clean water and air, social networks, access to health services, and now access to digital technologies[20] determine an individual's health status. Additionally, the interaction between these factors plays a role—an important interaction is the one between genetics and environment, whereby different environmental exposures affect people with different genes differently, or different genes determine the individual response to environmental factors.

According to Schleidgen et al.,[21] there are three main schools of thought around what precision medicine is:

1. Precision medicine is not a new concept as medicine has always been individualised.
2. Precision medicine is holistic healthcare centred around the needs of individual patients.
3. Precision medicine is treatment targeted to stratified subgroups.

In their systematic review, they found that the term "precision medicine" is used across different areas of medicine and acquires different meanings depending on who describes it. Nonetheless, improved tailoring of therapeutic interventions is a key characteristic of personalised medicine, as well as the use of molecular and genetic analysis to achieve personalisation. After screening over 2,000 research papers, Schleidgen et al. formulated the following definition of personalised medicine: "Personalised medicine seeks to improve stratification and timing of healthcare by utilizing biological information and biomarkers on the level of molecular disease pathways, genetics, proteomics as well as metabolomics." Interestingly, environmental factors are not mentioned, although proteomics and metabolomics could be seen as proxies for the environment which they are affected by.

However, despite the efforts to formulate a definition by synthesising the totality of studies available in the literature, De Grandis et al.[22] believe that this definition is limiting, as it focuses on personalised medicine as a developmental science and does not include aspects of patient-centred care such as clinical translation, implementation, and the idea that personalisation of care can also happen by considering the patient's needs holistically, and not just biologically. In other words, Schleidgen's definition of personalised medicine fails to include the fourth "P" of P4 medicine: participation.

As well as having different meanings for different people, personalised medicine is not free of scepticism. From the reluctance to accept that treatments work for one person but not another, to the belief that there are other, more pressing issues that need to be addressed before worrying about tuning drug doses depending on genetic variations, there are many counterarguments to the true value of personalised medicine. The causes of health disparity are not to be searched in our DNA, but in our environments; for instance, racism affects health, and more resources are needed to develop and evaluate interventions to combat the negative effects that racism has on health.[23] There is great concern that the focus on personalised medicine is diverting research funding away from tackling the healthcare divide driven by socio-economic factors.[24] Another source of health inequality within personalised medicine is the unaffordable pricing of the genetic tests required for pharmacogenomics. Accessibility to digital tools is yet another factor that can exacerbate inequalities, especially when introducing personalised medicine in countries where the socioeconomic divide is incredibly pronounced.

Despite significant progress made in human genome sequencing, characterisation of genetic variations, and identification of gene–drug interactions over the past 20 years, the field of precision medicine has not yet generated extensive clinical transformation. A big reason is that variation in treatment response is influenced not only by genetics but other biochemical characteristics, as well as environmental factors. Critics of personalised medicine, and in particular pharmacogenomics, believe that the assumption on which this discipline sits is inherently flawed: that the differential response to a drug is consistent for each individual, predictable, and based on stable properties, such as a genetic variant.[25] Pharmacogenomics essentially postulates that DNA alone is the cause of different treatment responses, and this can be a real shortcoming of the discipline. While this assumption may hold true for rare variants, complex conditions like CVD or T2D are highly polygenic[26] and are greatly affected by environmental factors driven by a sedentary lifestyle and unhealthy diet;[27] in these cases, it is not realistic to think that genetic information alone can inform on the suitability of a certain drug for a single individual.

N-of-1 trials aim to overcome the issues caused by varying responses to drugs. By taking measurements of the same person at multiple time points, researchers can adjust for those "non-fixed" factors, such as lifestyle and external environment, that may affect the measured outcomes. Clinicians routinely perform N-of-1 trials, for example, when optimising medications for patients with chronic conditions. N-of-1 trials are seen by many as the key to personalised medicine.[28,29,30,31] However, even the usefulness of N-of-1 trials for personalised medicine is not free from criticism, and some believe that N-of-1 trials are only useful if making strong assumptions, such as treatments not having long-term effects that may happen once the study is complete. Therefore, some believe that personalised medicine at the individual level will never be realised and that instead efforts should instead be focused on stratified medicine.[32] That is, defining and predicting more granular subgroups with differential treatment responses.

Considering the multitude of factors that affect human health is at the core of personalised medicine. Personalised medicine can be applied to any stage of the healthcare process: from more precise diagnoses, to individualised treatments and prevention strategies, and to prediction of disease risk. Regardless of any uncertainty around the definition of personalised medicine and its variants, the ultimate aim of personalisation is to overcome the one-size-fits-all approach that has been at the core of medicine so far. While population-based approaches surely have had, and continue to have, a great impact on human health, the future of medicine lies in personalisation.

Technology as the enabler of personalised medicine

Personalised medicine has, to an extent, always existed. Humoral theory was first applied to medicine by Hippocrates (460–370 BC), who, in his work "On the Nature of Man," described the human body as containing four humours: blood, phlegm, yellow bile, and black bile. Hippocrates describes the humoral theory as follows:

> The human body contains blood, phlegm, yellow bile, and black bile. These are the things that make up its constitution and cause its pains and health. Health is primarily that state in which these constituent substances are in the correct proportion to each other, both in strength and quantity, and are well mixed. Pain occurs when one of the substances presents either a deficiency or an excess, or is separated in the body and not mixed with others.[33]

Humoral medicine believed that a person's health is the combination of their humours, their lifestyle, and their environment;[34] there is, therefore, a clear connection between modern personalised medicine and the concept underpinning humoral medicine.

Even N-of-1 trials are not new: Mirza et al.[35] describe how, in the 18th century, a doctor in Bath reported using crossover periods in 13 patients to compare the purgative effects of three varieties of rhubarb, effectively performing a series of N-of-1 trials. Nowadays, clinicians routinely perform N-of-1 trials when trying to find the best drug and dosage for patients with chronic conditions.

Humoral theory remained at the core of Western medical practice until the mid-19th century, when empirical science took over.[36] Significant improvements in the microscope allowed two German scientists, Mathias Schleiden (1804–81) and Theodor Schwann (1810–82), to postulate cells as the building blocks of life.[37] With the help of the microscope, Louis Pasteur (1822–95), the father of microbiology, developed the germ theory of diseases, which states that illness is caused by external microorganisms infecting the body.[38] Medicine made great advancements in the 1800s and 1900s thanks to the invention of imaging technology such as the electrocardiograph, x-ray, and magnetic resonance imaging.

At the end of the 20th century, next-generation sequencing powered the genomic revolution that started when the first human genome was fully sequenced.[39] It also introduced a new category of data available for medical research—genomic data. Genomic data was followed by more categories that are now grouped under the -omics category, such as data from the transcriptome, epigenome, metabolome, etc. The availability of new data sources, as well as new computational tools able to process and analyse them, fast-started the modern idea of personalised medicine.

According to Hood et al.[40] the personalised medicine revolution is being powered by four factors:

1. An appreciation that medicine is an information science.
2. Systems or holistic approaches to studying the enormous complexities of disease.

3. Emerging technologies that will let us explore new dimensions of patient data space.
4. Powerful new analytical technologies that will let us interpret the multitude of data points associated with each individual.

Technology is a major tool powering the development of personalised medicine. Firstly, technological advancements have enabled the explosion of molecular data, fast-starting the *-omics* revolution. The term "omics" refers to the totality of biology disciplines that end in -omics, such as genomics, proteomics, metabolomics, epigenomics, and transcriptomics.[41] Genomics studies the structure, function, and mapping of genomes, while the transcriptome is the RNA expressed by a cell or tissue, providing a snapshot of cellular metabolism.[42] Epigenomics focuses on the study of the biochemical components and mechanisms that can activate or deactivate genes, controlling the production of a certain protein in particular cells. These three fields have made incredible advancements thanks to high-throughput next-generation DNA sequencing.[43] On the other hand, mass spectrometry allowed the widespread development of proteomics[44] and metabolomics,[45] which are concerned with the quantification of proteins and metabolites, respectively.

The capacity to measure the various -omics has led to the substantial and continuous expansion of research databases. In parallel, pivotal advancements in big data tools and enhanced computational capabilities have emerged as powerful allies to analyse and interpret these vast amounts of raw data. In particular, the evolution of big data technologies and machine learning methods has been instrumental. Innovations in data storage, retrieval, and processing have enabled researchers to efficiently harness the wealth of information contained within these datasets. Moreover, computational advancements such as high-performance computing and cloud-based infrastructure have enabled scientists to undertake analyses and simulations with previously unheard-of speed and accuracy. Machine learning techniques, especially deep learning, have enabled advancements in protein structure prediction and protein function prediction. For example, AlphaFold,[46] an AI system based on neural networks, is able to predict protein structures with unprecedented accuracy and speed. Using AlphaFold, it was possible to predict the structure of over 200 million proteins[47] that are now available for use in biomedical research.

New statistical and computational methods have also allowed the generation of new insights to advance epidemiology and our understanding of how genetic variants are associated with diseases. Genome-wide association studies (GWAS) are experimental studies that aim to identify the association of genotypes with phenotypes. Results from GWAS can lead to new insights into disease pathophysiology, or to the development of polygenic risk scores[48] (PRS) or genetic risk scores (GRS) that can be used as risk factors to predict an individual's likelihood of developing a disease.

GWAS have allowed uncovering the genetics underpinning thousands of diseases, including T2D and CVD. GWAS have identified over 100 common variants for T2D that have been replicated across major ethnic groups.[49] Results from GWAS on T2D have led to several improvements in medical research and evidence: pharma companies have developed new drugs; researchers have determined how vitamin D intake, early nutrition, circulating lipid levels, and chronic inflammation affect the development of T2D; PRS have been developed as valuable tools to improve disease stratification and prognostic models.[50] For CVD, PRS have been found to provide a better indication of coronary artery disease risk compared to six risk factors conventionally used in clinical practice (smoking, diabetes, family history of heart disease, body mass index, hypertension, Hypercholesterolaemia).[51]

Mendelian randomisation (MR) is another statistical technique that has allowed the translation of results from GWAS into useful medical information. MR uses genetic data to uncover causal associations between risk factors and diseases by using genetic variants strongly associated with these factors as proxy.[52] MR has, for example, confirmed the causal role of physical activity in lowering diastolic blood pressure and in decreasing the risk of hypertension.[53]

Computational techniques like GWAS and MR rely on very large sample sizes as well as a plethora of data around lifestyle, environment, health, biomarkers, etc., in order to uncover useful genetic variants. The introduction of biobanks has been crucial to these discoveries, as they provide data from thousands of genotyped individuals for whom detailed data was collected through questionnaires, laboratory measurements, and electronic health records. The UK Biobank is a significant example of how biobanks have enabled GWAS of hundreds of traits and diseases, such as chronic pain,[54] depression,[55] physical activity,[56] osteoarthritis,[57] and many more.

Personalised medicine is not only realised through biomedical discovery but is also enabled by new technologies introduced within clinical practice. There are several ways in which technology continues to enable the implementation of personalised healthcare. The most notable example is, arguably, digital health. First introduced in 2000 by Seth Frank,[58] digital health encompasses several technologies, including mobile health, health information technology, wearable devices, telehealth, telemedicine, and artificial intelligence. Digital health has three broad applications:[59] improving system efficiency, informing and delivering monitoring to encourage behaviour change and patient self-management, and aiding diagnosis and guiding treatment. For example, telemedicine is a way of delivering medical care over the internet, using video chat, messaging, or other applications, providing better access to care—especially if patients are housebound or live in remote areas—and optimising resource management.[60] Telemedicine applications span several use cases that are not limited to patient consultations and can include remote monitoring, telehealth nursing, remote physical and psychiatry rehabilitation. Clinical decision support systems (CDSS) are another example of digital health application in clinical practice. CDSS are a direct aid to clinical decision-making, usually realised by matching individual patient information to a clinical knowledge base to issue recommendations that the clinician can use for decision.[61] For this, CDSS often use information held in electronic health records, another digital technology that revolutionised clinical practice.[62] CDSS have several applications, including diagnostics, alarm systems, disease management, prescriptions, and drug control.[63]

PERSONALISED NUTRITION AND PHYSICAL ACTIVITY

Nutrition, physical activity, and lifestyle in general are well-recognised contributors to health status, disease prevention, and life expectancy.[64] Common diseases such as T2D, CVD, and some forms of cancer have been closely linked with nutrition,[65] physical activity,[66] and sedentary behaviour.[67]

Similar to personalised medicine, personalised nutrition and physical activity aim to identify person-specific features able to explain and predict the individual response to lifestyle factors, such as particular foods, diets, and exercise regimens. While there is a large overlap between personalised medicine and pharmacogenomics, or optimisation of medical therapies based on personal factors, other areas of personalised medicine focus on personalising lifestyle interventions in order to maximise their benefits. When talking about personalised, or precision, nutrition and physical activity, we refer to research showing how individuals respond differently to different diets, but also how exercise modalities and nutrients affect the risk of lifestyle-related conditions depending on individual factors, including genetics. With the knowledge that individuals respond differently to foods, personalised nutrition involves, for example, designing dietary plans specifically suited to an individual's unique biological makeup, preferences, and psychological characteristics. On the other hand, personalised physical activity aims to create individualised exercise regimens based on factors such as genetic predispositions, metabolic profiles, and fitness goals. By integrating personalised physical activity and nutrition into the broader framework of personalised healthcare, practitioners can design tailored plans that address the unique characteristics and needs of each individual.

Personalising nutrition

Dietary interventions are routinely used due to their efficacy in preventing and managing conditions triggered by a poor lifestyle. T2D is a notable example of a disease for which nutrition therapy is an effective component to manage the condition. Medical nutritional therapy can achieve the same improvements in haemoglobin A1c (HbA1c) as medication for T2D, and it can help decrease the dose of glucose-lowering medications, as well as improve the self-perception of health status, motivation, and emotional stress.[68] However, a one-size-fits-all approach is an unrealistic expectation given the broad spectrum of people with diabetes and prediabetes, their cultural backgrounds, personal preferences, comorbidities, and socioeconomic environment;[69] and despite the known link between diet and disease, dietary interventions to improve public health and wellbeing have had limited impact.[70] As a result, a consensus report published by the American Diabetes Association[71] suggests the use of individualised plans for nutrition therapy in T2D to tailor nutritional therapies to dietary preferences, resources, food availability, the individual's health literacy, and other factors. T2D and other lifestyle-related conditions are highly heterogeneous as they are characterised by different genetic, pathophysiological, and clinical profiles[72] and as such, one-size-fits-all approaches to prevent and manage these conditions, although useful, may have limited efficacy. Personalised approaches have the potential to improve outcomes by considering individual factors, rather than the average response at the population level.

Personalised nutrition does not have an agreed definition. Ordovas et al.[73] define it as "an approach that uses information on individual characteristics to develop targeted nutritional advice, products, or services," which overlaps with related concepts such as nutrigenomics, nutrigenetics, and nutritional genomics. However, in the same way that personalised medicine expands further than pharmacogenomics, personalised nutrition is not limited to employing genetic information to tailor dietary interventions. It can make use of phenotypic, medical, nutritional, and other data to deliver more precise nutritional guidance. Personalisation can also be based on behavioural factors and food preferences, as well as environmental and social factors. Recently, more advanced forms of personalisation have been made possible by more advanced technologies, particularly around -omics disciplines. Genomic discoveries have enabled the emergence of nutrigenetics, a discipline that examines how genotypes influence the individual response to diet, as well as exploring how diets modulate the relationship between genetics and disease. Advances in molecular and computational methods have uncovered the role of the gut microbiome in determining how different individuals respond to different foods.

Many GWAS have identified genetic variants associated with T2D and glucose metabolism,[74] as well as variants associated with macronutrient intake.[75,76] These genetic markers can inform who is most likely to benefit from a certain diet. Research has been carried out to explore the effect of single nucleotide polymorphisms (SNPs) on the individual response to different diets, as well as elucidating the modulating effects that diet can have on the risk of disease associated with these polymorphisms. Most of the research on gene–diet interactions has been conducted in relation to diseases associated with lifestyle, such as obesity and T2D. As a result, there is extensive evidence that genetics and diet interact to affect fasting glucose, insulin resistance, and weight loss. For example, several SNPs can inform the efficacy of a ketogenic diet,[77] which is low in carbohydrates and high in fats. The effect allele in rs814628, a SNP in the gastric lipase gene involved in lipid digestion, is associated with more pronounced weight loss in response to a ketogenic diet.[78] Similar evidence is also available for SNP rs5883, found in the gene for cholesteryl ester transfer protein, which regulates the transport of cholesterol from peripheral tissues to the liver, suggesting that an individual's ability to metabolise lipoproteins affects the extent of their weight loss.[79]

Blanco-Rojo et al.[80] found that variants in the S100 calcium-binding protein A9 that are associated with higher risk of T2D are also associated with higher insulin resistance following a diet with a high saturated fats to carbohydrate ratio. However, when the fats:carbohydrate ratio was low, measures of insulin resistance were comparable to those of people without increased T2D risk. This finding shows that people with the high-risk genotype could benefit from a low saturated fats:carbohydrate ratio intake to improve insulin resistance. Importantly, the results have been confirmed in three populations of different ancestries: white Spanish, North American non-Hispanic, and Hispanic.

Another genetic variant associated with dietary intake was reported by Kang et al.,[81] who showed that diabetic individuals carrying the risk allele in the apolipoprotein A5 gene benefit from replacing refined rice with whole grains and legumes in a high-carbohydrate diet, as this prevents diabetic hypertriglyceridemia.

Hjorth et al.[82] used fasting plasma glucose and fasting insulin to predict whether diets with different glycaemic loads or fibre and whole-grain content would lead to weight loss and maintenance in different individuals. Prediabetic individuals were more susceptible to weight regain and weight loss, and responded better to a high-fat and low-carbohydrate diet, or one high in fibre and whole grain. These results show that individual glucose and insulin levels can be used to choose the most optimal diet.

Meta-analyses conducted by Dashti et al.[83] confirmed that reducing carbohydrate intake can reduce the negative effects of genetic variants in the melatonin receptor 1B associated with glycaemic traits, namely fasting glucose and Homeostatic Model Assessment for Insulin Resistance (HOMA-IR). Specifically, for individuals with the T allele in the rs1387153 variant, the association between carbohydrate intake and fasting glucose is significant, and therefore people with this variant should reduce carbohydrate and increase monounsaturated fatty acids intake.

Variants in the PSMD3,[84] PCSK7,[85] IRS1[86] genes are associated with T2D and modify the effect of diet on insulin resistance.

Conditions like T2D and obesity are highly complex, and multiple genes contribute to the risk of developing the conditions. SNPs typically have small effect sizes and low predictive power when used to infer disease risk.[87] As such, extensive research exists to uncover the interaction between disease, diet, and PRS, genetic risk scores that are calculated based on multiple genetic variants, usually SNPs. PRS have been found to explain more variance in phenotypes compared to SNPs, which is due to single variants having very small effects, as well as causal variants being rare and therefore harder to detect.[88] In the context of T2D, Huang et al.[89] found that macronutrient intake modifies the genetic association with T2D, using a PRS calculated from 31 variants associated with the condition. In particular, low protein intake improved insulin resistance and reduced HbA1c to a higher extent in individuals with a low genetic score compared to higher genetic scores, suggesting that the level of protein in a diet aimed at weight loss in patients with T2D, hypertension, or hyperlipidaemia should be modulated based on genetic factors.

Villegas et al.[90] explored the effect of fibre intake on the relationship between a PRS calculated from 15 SNPs and T2D. They found that carbohydrate and fibre intake modify the genotype–phenotype relationship differently based on ethnicity.

Similar evidence is also available for CVD, and several studies have shown that the link between SNPs and cholesterol—one of the most significant risk factors for CVD—is modified by fat intake,[91,92] alcohol intake,[93] and carbohydrate intake.[94] Diet can also modify the genetic risk of cardiovascular conditions, as alcohol consumption,[95] coffee intake,[96] saturated fat intake,[97,98] vegetable intake,[99] and protein intake,[100] have been found to interact with SNPs to affect the risk of myocardial infarction. Significant gene–environment interactions have also been shown for CVD[101,102] and coronary heart disease.[103,104]

Information from GWAS and other association studies considering SNPs and PRS can be used to develop personalised interventions that can improve health outcomes more effectively. The efficacy of dietary advice based on genetic information has been tested in a few studies. A very early example of how genetic information can personalise dieting is the study by Arkadianos et al.,[105] who showed that a diet personalised with 24 genetic variants improves body mass index (BMI) and glucose control in people with a history of weight loss failure. The result was more recently replicated by Vranceanu et al.,[106] in which a personalised DNA-based diet leads to higher weight loss at 18 months than a ketogenic diet. The DNA-based diet also improved glucose metabolism, triglycerides, and high-density lipoprotein cholesterol (HDL-C). More recently, Karvela et al.[107] showed that, in individuals with non-diabetic hyperglycaemia, personalisation of dietary advice based on DNA was associated with a significant reduction in fasting plasma glucose and HbA1c at 26 weeks when compared to standard care.

The microbiome, which refers to the collective genomes of the microbiota that inhabit the human gut, regulates food digestion and absorption, energy harvesting, and other aspects of metabolism, and plays a role in the success of weight loss interventions. Emerging evidence suggests that differences in the gut microbiome affect individual responses to dietary interventions.[108] The amount of specific bacteria may indicate how an individual responds to a certain diet. For example, healthy individuals whose glucose metabolism improved after consuming barley kernel-based bread had a higher ratio of *Prevotella/Bacteroids* compared to non-responders, suggesting that these bacteria can promote increased glycogen storage.[109]

Differences in the gut microbiome can explain individual differences in postprandial glucose response (PPGR) to the same foods. Zeevi et al[110] found significant associations between PPGRs to standardised meals and both clinical and microbiome data: they found HbA1c, BMI, systolic blood pressure, and alanine aminotransferase activity positively associated with PPGRs. Most notably, they found associations between PPGR and microbiome features, showing that PPGRs vary greatly across different individuals and are associated with both person-specific clinical and microbiome factors. They also used these insights to show that a dietary intervention personalised on microbiome, blood parameters, continuous glucose monitoring, anthropometrics, and various questionnaires leads to lower PPGR, as well as fewer glucose spikes and fluctuations. The benefits of a diet tailored to PPGR based on individuals' clinical and microbiome features were replicated in a small cohort of adults newly diagnosed with T2D, for whom the personalised dietary intervention led to significantly lower levels of glucose compared with a commonly recommended Mediterranean-style diet.[111]

Despite promising early results, there is a need to gather more evidence before introducing nutrigenomics, and personalised therapies in general, within clinical practice. Although extensive research has proved that gene–diet interactions exist for lifestyle-related conditions, there is little consistency in the methods used by the various studies. Studies have used a wide range of dietary measures, follow-up times, genetic measures (SNPs, weighted GRS, unweighted GRS, etc.), which makes it challenging to compare them and collate them into a systematic body of evidence.[112]

Additionally, most studies that examine the interaction between diet and microbiome show associative relationships, rather than causative, and rarely describe the possible mechanisms underlying the associations.[113] Evidence is not consistent, and despite an important number of studies showing significant associations between diet and genetic variants, Berry et al.[114] have found that genetics influenced individual food responses less than expected, while other factors such as meal timing and composition had a bigger role. Lack of conclusive evidence is not the only barrier to adoption—conflicting results, scepticism in the healthcare community, and the high cost of -omics technologies also prevent personalised medicine from being introduced in medical practice.

Despite significant challenges, the mounting body of evidence suggests that personalised nutrition may have the potential to significantly improve health, and efforts are being made to collate the existing evidence into guidelines and frameworks to encourage the adoption of personalised nutrition in clinical practice. Firstly, this will be achieved by standardising research methods such that studies can be compared across populations and results summarised in systematic reviews. Meta-analyses will be particularly important in creating the evidence necessary to introduce personalised nutrition at the point of care. Secondly, epidemiological analyses will be needed to prove the utility of personalised diets at the population level.[115] Thirdly, frameworks are needed in order to evaluate interventions consistently. Grimaldi et al.[116] proposed a framework to assess the quality of evidence around personalised nutrition, which suggests considering study design and quality, as well as biological plausibility of the interaction, to assess the scientific validity of interactions. Results from the Food4Me study, a multi-country RCT assessing the efficacy of personalised diets, show that knowledge of one's genetic risk does not lead to better outcomes compared to advice personalised using the individual's phenotypic characteristics,[117,118] even though fat mass and obesity associated (FTO) risk carriers had a greater reduction in obesity-related traits following the intervention.[119]

Finally, as more and more companies offer direct-to-consumer testing and personalised advice, some healthcare professionals are starting to consider such evidence when developing nutritional therapies for their patients. The drive to incorporate nutrigenomics into real-world practice has been formalised by Horne et al.,[120] who developed the first nutrigenomic care map. However, more work is needed to support clinicians in leveraging nutrigenomics, and some argue that it is imperative to create an evidence-based data repository, informatic platforms, and educational resources for clinicians in order to facilitate adoption.[121]

Personalising physical activity

Physical activity, defined as "any bodily movement produced by skeletal muscles that requires energy expenditure,"[122] has a well-documented impact on health and wellbeing. Physical activity significantly reduces the risk of cancer, T2D, ischaemic heart disease, and stoke events[123] and has a dose–response relationship with health outcomes, meaning that while any level of activity is better than none, the more time someone spends active, the more their risk of disease is reduced.

As the evidence on the benefits of physical activity is clear, the World Health Organization (WHO) recommends a minimum of 150 minutes per week of moderate-intensity physical activity, or 75 minutes of vigorous physical activity.[124] However, the current guidelines do not specify the type, duration, and intensity of physical activity, despite there being evidence that different types of physical activity affect biomarkers of health in different ways.[125] Several studies have documented the benefits of LIPA on insulin sensitivity,[126] hyperglycaemia,[127] postprandial glucose,[128] cardiometabolic risk factors,[129] and all-cause mortality.[130]

In the last few years, the negative effects of sedentary behaviour have also been brought to light. Sedentary behaviour is defined as any activity below 1.5 metabolic equivalent of task in either a lying, reclining, or sitting position,[131] and it increases the risk of all-cause, CVD, and cancer mortality, as well as incidence of CVD, cancer, and T2D.[132] Sedentary behaviour has an important impact on T2D, as it is unfavourably associated with cardiovascular and metabolic biomarkers[133,134] and with glycaemic control[135] independently of physical activity. As a result, the WHO[136] and countries such as the UK[137] and Australia,[138] have started introducing guidelines on sedentary behaviour. While guidelines do not specify how to reduce the time spent in sedentary behaviours, there is evidence of a dose–response relationship between sedentary behaviour and all-cause, CVD and cancer mortality, and incident CVD, and that replacing sedentary time with activity of any intensity can counteract these negative effects.[139] Exercise is not the only way to reduce the health risks of a sedentary lifestyle, and even low levels of activity have been found to effectively counteract negative effects of sedentary time.[140]

Given the important influence that physical activity has on health, exercise is a crucial part of lifestyle management interventions, which are recommended as the first-line therapy for many conditions triggered by an unhealthy lifestyle. For example, physical activity is the first-line therapy to improve glucose control in T2D.[141]

But similar to nutrition advice, guidelines on physical activity are modelled on the whole population, with little regard for individual differences, needs, and abilities. Even more so than for nutrition advice, the current public health guidelines on physical activity and sedentary behaviour have a truly one-size-fits-all format and are largely general—they give no indication of the type or intensity of the activity, and they do not specify how to reduce the time spent on sedentary behaviours. And while it is deemed that the current evidence is not sufficient to support more specific guidelines,[142] several studies highlight how the different individual responses to exercise may affect the relationship between physical behaviours and health.

Everyday life experience, as well as a plethora of research studies, shows us that individuals respond very differently to regular physical training, be it for weight loss, improved physical performance, or reduction of negative health outcomes.[143] The first study exploring the variation in response to training was published in 1984,[144] and since then many studies have shown that people respond differently to exercise by showing different changes in maximum oxygen uptake (VO_2max), which is the standard parameter of cardiorespiratory fitness, as well as disease-specific parameters such as fasting glucose, insulin sensitivity, or triglyceride levels. Additionally, it has been observed that some individuals do not seem to respond to training,[145] although there is scepticism around the real existence of non-responders.[146,147] However, a recent study[148] showed that increasing exercise intensity does not improve glycaemic response in people with prediabetes and T2D. The authors identified non-responders based on changes in HbA1c following 16 weeks of aerobic exercise and administered a further 12 weeks of increased intensity training; most initial non-responders did not display improvements after the second round of training.

Some individuals have an adverse response to training. A review by Bouchard et al.[149] showed that adverse metabolic responses to exercise are quite common, with around 10% of adverse responders in six exercise training studies, as they showed worse blood pressure, higher triglycerides, higher fasting insulin, and worse cholesterol after completing the training course. To prevent adverse reactions, there is a clinical need to predict subjects who do not respond well to exercise, especially as these adverse responses are quite prevalent. These adverse reactions do not include those caused by rare conditions, such as sudden cardiac death due to cardiomyopathies, but are due to physiological mechanisms that are still not fully understood.

In summary, the individual response to exercise can be challenging to characterise. Individuals who do not respond to one exercise modality may respond to others,[150] and some may take longer to respond to exercise[151] or may need higher volumes to respond.[152] Non-response can be addressed by creating individualised plans to adjust exercise modality, intensity, and duration. Given that nutrition, stress, and sleep quality affect response to training,[153] there may also be variation within the same person, although there is no conclusive evidence showing this. Exercise interventions can also be designed to respond to variations within the individual, for example by using heart rate variability to modulate the intensity of the training sessions.[154]

The mechanisms underlying the different responses to exercise are not yet fully understood, partially owing to the great number and complexity of the factors affecting response to exercise. In general, age, sex, race, and initial fitness level have little influence on response to standardised training.[155] Factors such as baseline phenotypes, training history, nutritional intake, psycho-emotional states, between-session recovery, weight, and ability to meet prescribed training workloads play a larger role in determining how someone will respond to exercise.[156] The role of genetic factors in particular has received a lot of interest as it is believed to have a significant role in determining how a person will respond to the type of exercise. For example, an SNP in the ACTN3 gene that regulates the expression of a protein which constitutes a major structural component of skeletal muscles was found to be linked with greater knee extensor peak power response to strength training.[157] Rankinen et al. found that nine SNPs explained approximately 36% of the variation in heart rate induced by a 20-week endurance training program.[158] Additionally, improvements in VO_2max following exercise interventions were found to be associated with several SNPs.[159] Genetic associations have also been found for time-to-completion of Ironman events and with gains in endurance fitness following exercise training.[160] However, genetics can only partially explain individual variations. For example, Sarzynki et al. found that genetic heritability only accounted for less than 50% of the variation in VO_2max[161] in response to standardised exercise training programs. Another study found that while genetics explains cross-sectional differences in traits, it did not play a big role in explaining the response to exercise, and that environmental factors are more important.[162]

Considering the interaction between genetics and lifestyle can help uncover a higher portion of the unexplained variance in traits associated with exercise response. Several gene–lifestyle interactions have been reported for physical behaviours. For example, exposure to elevated sedentary levels and high intake of calories, sugar, and fat increase the genetic susceptibility to higher BMI,[163] and the negative effects of sitting have a greater impact on individuals with the risk allele in the FTO gene, which is also known as the obesity gene.[164] In a prospective cohort study,[165] the reduction in T2D risk afforded by physical activity was more pronounced in individuals with a low genetic risk, with the strength of the association between T2D and activity being reduced in individuals with a high genetic risk.

A large amount of evidence shows the impact of gene–environment interactions on serum lipid levels. Higher levels of physical activity enhance the high-density lipoprotein-increasing effects of certain loci and attenuate the low-density lipoprotein cholesterol-increasing effect of others, thus decreasing the risk of CVD.[166] Higashibata et al. showed that leisure-time physical activity is associated with lower triglyceride levels depending on the genotype in the NOS3 gene, which encodes the production of nitric oxide, a molecule expressed in the endothelial tissue that lines blood vessels.[167] An intervention study found that postmenopausal women with the minor allele in the same SNP did not display reductions in triglycerides following exercise training.[168] The relationship between HDL-C and hepatic lipase, an enzyme that catalyses the hydrolysis of triglycerides and phospholipids, is modulated by physical activity such that vigorous physical activity was associated with improved cholesterol only in homozygous A-allele carriers.[169]

The practical applications of exercise genomics are ubiquitous, from creating individualised exercise plans to predicting adverse responders. However, despite mounting proof of the effect of genetics and other individual factors on exercise variability, the quality of the current evidence is not high enough to be actionable, and more intervention studies with a control design are needed.[170]

TYPE 2 DIABETES AND PERSONALISED MEDICINE

Globally, over 6% of the world's population is affected by T2D, making this disease the ninth leading cause of mortality and the seventh leading cause of disability-adjusted life years.[171] T2D prevalence is expected to rise from just over 6,000 per 100,000 in 2017, to almost 8,000 per 100,000 by 2040.[172]

Epidemiological studies have highlighted how behaviour, lifestyle, and biological risk factors impact the likelihood of developing the condition. Increased adiposity is the most important factor, and diet plays a crucial role in determining T2D risk.[173] Physical activity and sedentary behaviour are also particularly important risk factors, as described thoroughly in earlier parts of this chapter. Other behaviours associated with T2D are smoking,[174] sleep,[175] television-watching,[176] and shift work.[177] Additionally, T2D is associated with socioeconomic development, both on a macro scale, with more developed countries exhibiting higher prevalence rates,[178] and on a micro scale, with lower individual socioeconomic status associated with higher risk of developing the condition.[179]

Rising numbers of T2D cases are attributable to obesity, sedentary lifestyles, and ageing populations. T2D is associated with a higher risk of other conditions, including coronary heart disease, ischaemic stroke, and other CVDs. Therefore, it is not surprising that lifestyle management is the first line of therapy for people with T2D and for those with an increased risk of developing the condition. In fact, T2D can be mitigated by managing obesity and impaired glucose regulation with diet and exercise interventions more than with drug therapies.[180] Nonetheless, despite lifestyle interventions being initially successful, poor adherence in the long term leads to weight regain, and therefore most patients with T2D are prescribed drug treatments, predominantly pioglitazone, metformin, and rosiglitazone.[181] However, up to 90% of cases can be prevented by following a healthy diet, maintaining a healthy BMI, exercising regularly, and avoiding smoking and excessive alcohol consumption.[182]

Pathophysiology

T2D is characterised by insulin deficiency caused by insulin resistance in target organs, leading to pancreatic beta-cell dysfunction and consequent loss of insulin secretion.[183] The pancreas, liver, skeletal muscle, kidneys, brain, small intestine, and adipose tissue are the organs involved in the disease. T2D is preceded by prediabetes, a condition in which fasting glucose levels and glucose tolerance are impaired, and HbA1c levels are increased.

While insulin is central to T2D, other proteins such as incretin hormones, adiponectin, ghrelin, and leptin play an important role in the disease. The incretin hormones glucose-dependent insulinotropic polypeptide (GIP) and glucagon-like peptide-1 (GLP-1) are gut proteins secreted after food intake that stimulate insulin production (i.e., the incretin effect). In individuals with T2D, the incretin effect is impaired despite secreted levels being comparable to healthy subject.[184] Given its ability to reduce hyperglycaemia, GLP-1 is considered to have promising therapeutic potential for the treatment of obesity and T2D.[185]

Adiponectin is a hormone with anti-inflammatory and insulin-sensitising properties and is associated with lower T2D risk, marking this molecule as an important hormone related to T2D.[186] Inflammatory markers, such as C-reactive protein (CRP), interleukin-6, and tumour necrosis factors, are associated with increased risk of T2D, making chronic inflammation a strong predictor of T2D development.[187]

Ghrelin is a hormone with insulin-suppressing properties, and therefore, blocking its signalling is considered a promising option for T2D treatment.[188] While ghrelin is commonly known as the "hunger hormone," leptin, another hormone involved in metabolism regulation, is also termed "the satiety hormone." Leptin is a small protein involved in metabolism and body weight regulation whose deficiency results in obesity and T2D.[189] Leptin is produced by the adipose tissue, and its receptors are expressed mainly in the hypothalamus, where leptin affects the dopamine pathway responsible for appetite and food intake.[190] Leptin has been proposed for use in therapies for leptin-deficient patients[191] and is a potential biomarker in obesity management therapies, as leptin resistance leads to obesity[192] and leptin levels vary in relation to dieting therapies.[193]

Additionally, leptin is a biomarker for estimating the risk of ischaemic heart disease in people with T2D,[194] and for differentiating between Crohn's disease and ulcerative colitis.[195]

High leptin levels are associated with insulin resistance and the development of T2D. Moreover, in people with T2D, hyperleptinemia is correlated with an increased risk of obesity and cardiovascular outcomes.[196] Using leptin as a biomarker can have research and clinical significance, as rapid detection methods can accelerate research into the physiological mechanisms through which leptin regulates energy balance and immune response, which still remain poorly understood.[197] Moreover, given that leptin levels change at a greater rate compared to weight loss under an energy-restricted diet[198] monitoring leptin concentration during dieting therapies can provide personalised insights on the success of the intervention.

Complications of diabetes

Cardiovascular conditions are complications traditionally associated with T2D. On the macrovascular scale, they include coronary heart disease, stroke, and peripheral arterial disease, and on the microvascular front, they include diabetic kidney disease, retinopathy, nephropathy, and peripheral neuropathy.[199] T2D is an important risk factor for the development of CVD, with individuals affected having a two- to four-fold higher risk of developing cardiovascular problems.[200] Additionally, patients with T2D share many of the risk factors for CVD, namely obesity, dyslipidaemia, and hypertension.[201] Lifestyle factors associated with T2D incidence, such as alcohol consumption, physical inactivity, sedentary behaviour, smoking, and unhealthy diets, are also risk factors for CVD and cardiovascular complications in T2D—individuals with T2D and a very healthy lifestyle have over 50% lower risk of CVD compared to those with the least healthy lifestyle.[202]

CVDs have historically been the leading causes of morbidity and mortality in individuals with T2D, but studies from the past ten years indicate that cancer is predicted to become the leading cause of diabetes-related death.[203]

Diabetes and genetics

T2D is characterised by a significant heritable genetic component—individuals with a sibling who has the disease have a two- to three-fold increased risk of developing the condition.[204] T2D is a highly complex and heterogeneous disease that is associated with hundreds of genetic variants, although they only increase the disease risk by 10–20%.[205]

Given that the majority of heritability cannot be explained by genetics alone, gene–environment interactions are thought to be particularly important in determining the risk of T2D, with certain variants increasing the susceptibility to an environment characterised by sedentary behaviour and high consumption of sugar and fat. There is strong evidence of the interaction of the FTO gene, which is strongly correlated with BMI and T2D, with physical activity and sedentary behaviour.[206] As discussed previously, a large amount of evidence shows the impact of the interactions between genetics and physical activity on serum lipid levels.

Several studies have documented the effects of the interaction between genetics and diet on diabetes. Reducing carbohydrate intake can reduce the negative effects of genetic variants in the melatonin receptor 1B associated with fasting glucose and HOMA-IR,[207] which are biomarkers of diabetes. Additionally, carbohydrate and fibre intake were found to modify the genotype–phenotype relationship between 15 SNPs and the risk of T2D.[208] Other studies found variants in the PSMD3,[209] PCSK7,[210] and IRS1[211] genes to be associated with T2D and modify the effect of diet on insulin resistance.

Lifestyle behaviours and diabetes

As described in previous sections, physical activity is critical to manage blood glucose and improve overall health in people with T2D and prediabetes.[212] Adults with T2D should exercise at least every other day for at least 150 minutes per week, reduce sitting time, and break prolonged sitting bouts with light activity every 30 minutes. Sedentary behaviour—any waking activity with low energy expenditure—is unfavourably associated with several T2D biomarkers and the risk of all-cause, cardiovascular, and cancer mortality.[213] Additionally, replacing sedentary time with physical activity and breaking prolonged bouts of sitting reduces BMI, waist circumference (WC), and triglycerides, and increases HDL-C.[214, 215]

Individuals with impaired glucose metabolism are advised to change their dietary habits as there is strong evidence that nutritional therapy benefits the prevention and management of T2D. Medical nutritional therapy can lower HbA1c in similar proportions to medication, and the weight loss induced by changes in nutritional patterns can delay T2D by over three years.[216]

Chapter 2 thoroughly discusses the role of physical behaviours in individuals with T2D.

CASE STUDY: DATA DRIVEN IDENTIFICATION OF TYPE 2 DIABETES PHENOTYPES

T2D is a highly heterogeneous disease, characterised by different clinical profiles and therapeutic responses. Medically recognised examples of such heterogeneity are latent autoimmune diabetes in adults, gestational diabetes, and keratosis-prone T2D.[217] In order to find whether other sub-types of diabetes exist, Ahlqvist et al.[218] clustered diabetes patients based on six variables—glutamate decarboxylase antibodies, age at diagnosis, BMI, HbA1c, and homeostatic model assessment 2 estimates of β-cell function and insulin resistance. Excluding type 1 diabetes, they found four distinct clusters: severe autoimmune diabetes, characterised by early onset, low BMI, poor metabolic control, insulin deficiency, and with glutamic acid decarboxylase antibodies (GADA); severe insulin-deficient diabetes, similar to the first cluster but without GADA; severe insulin-resistant diabetes, characterised by high insulin resistance and high BMI; mild obesity-related diabetes, including patients with obesity and no insulin resistance; and mild age-related diabetes, with higher onset age and modest metabolic derangement. Moreover, the study found different disease progression between the clusters—especially for kidney-related diseases—and different genetic markers associated with each cluster.

Several other studies have explored the clustering of diabetic patients[219,220,221,222,223] and have analysed the clusters in terms of disease risk and progression. The clusters were determined with various methods, including K-means,[224,225] hierarchical clustering,[226,227,228] TwoStep clustering,[229,230] and using features including blood chemistry, anthropometric data, and measures of glucose metabolism. Dennis et al.[231] discussed the clinical relevance of the clusters in terms of differential drug treatments that can be administered based on the characteristics of the cluster and the risk of complications. Li et al.[232] explored associations with diabetes complications, Kahkoska et al.[233] assessed the risk of major cardiovascular events, while Karpati et al.[234] evaluated the different disease progression trends between the clusters. A complete description of the existing works on diabetes clusters is provided in Table 1.1.

The case study presented here addresses the following research question: given that clusters in diabetes have been found in several populations, is it possible to reproduce the result in the UK population, using cardiometabolic variables that are more commonly found in observational studies and more readily available in clinical practice?

The UK Biobank

The UK Biobank is a large-scale biomedical database and research resource, containing in-depth health and genetic information from half a million UK participants. Over 500,000 adults aged 37–73 were recruited in the UK between 2006 and 2010, with several follow-ups conducted in later years to collect more comprehensive genetic data or to collect data from other sources, for example accelerometer. Participants provided detailed information about themselves through questionnaires, interviews, and various physical measurements, along with samples of blood, urine, and saliva. Furthermore, the UK Biobank is enhanced by ongoing collection of data through health records, including GP records, hospital visits, and cancer and death registries. The study design of the UK Biobank is described in detail by Sudlow et al.[235] Ethics approval for the UK Biobank study was obtained from the North West Centre for Research Ethics Committee (REC reference: 21/NW/0157). Informed consent for the UK Biobank study was obtained from participants during the baseline assessment. The work described here was approved by the UK Biobank under application ID 56132. All analyses were performed in accordance with the relevant guidelines and regulations.

At recruitment, participants completed a questionnaire and verbal interview about sociodemographic factors, family history, early life exposures, psychosocial factors, environmental factors, lifestyle, and health status. At the same baseline visit, they had physical measurements taken and blood, urine, and saliva samples

TABLE 1.1
Summary of the studies analysing clusters of diabetes

Ref.	Method	Features for clustering	Analysis on clusters
Ahlqvist et al.[a]	TwoStep clustering	Glutamate decarboxylase antibodies, age at diagnosis, BMI, HbA1c, and HOMA estimates of β-cell function and insulin resistance.	Cox regression and associations with genetic markers.
Anjana et al.[b]	K-means	Age at diagnosis, BMI, waist circumference, HbA1c, triglycerides, HDL-C, and C peptide.	Cox regression (proportional hazards).
Cho et al.[c]	Hierarchical clustering	Age, gender, BMI, hypertension, and family history of diabetes.	Differences in biochemical profiles and performance of risk prediction.
Zaharia et al.[d]	Model trained by Ahlqvist et al.	Glutamate decarboxylase antibodies, age at diagnosis, BMI, HbA1c, and HOMA estimates of β-cell function and insulin resistance.	Progression of metabolic variables, liver steatosis/ fibrosis, nephropathy, and neuropathy over five years.
Li et al.[e]	TDA-based clustering	73 features including blood count, blood chemistry, and anthropometric data.	Associations with complications and genetic markers.
Kahkoska et al.[f]	Model trained by Ahlqvist et al.	Glutamate decarboxylase antibodies, age at diagnosis, BMI, HbA1c, and HOMA estimates of β-cell function and insulin resistance.	Risk of major cardiovascular events, cardiovascular death, all-cause death, heart failure, nephropathy, hypoglycaemia.
Karpati et al.[g]	Trajectory clustering and K-means	24 features derived from HbA1c trajectories.	Disease prevalence and progression.
Arevalo Lorido et al.[h]	Hierarchical clustering	Obstructive pulmonary disease, dyslipidaemia, liver disease, dementia, stroke, arrhythmias, systolic blood pressure, BMI, eGFR and haemoglobin.	Description of clusters in terms of the features.
Alexandre et al.[i]	Hierarchical clustering	Diabetes self-management behaviours (healthy eating, physical activity, self-monitoring of glucose, and foot care), self-efficacy (disease management), empowerment, distress, quality of life.	Differences between clusters in terms of sociodemographic, clinical, and care delivery process variables.

(*Continued*)

TABLE 1.1 (CONTINUED)
Summary of the studies analysing clusters of diabetes

Ref.	Method	Features for clustering	Analysis on clusters
Dennis et al.[j]	Model trained by Ahlqvist et al.	Glutamate decarboxylase antibodies, age at diagnosis, BMI, HbA1c, and HOMA estimates of β-cell function and insulin resistance.	Five-year progression in HbA1c change, kidney disease, albuminuria. Differences by cluster in treatment response.
Amato et al.[k]	TwoStep clustering	GLP-1, GIP, ghrelin.	Differences in clinical and anthropometric characteristics between clusters.

[a] Ahlqvist et al., "Novel Subgroups of Adult-Onset Diabetes and Their Association with Outcomes: A Data-Driven Cluster Analysis of Six Variables."

[b] Anjana et al., "Novel Subgroups of Type 2 Diabetes and Their Association with Microvascular Outcomes in an Asian Indian Population: A Data-Driven Cluster Analysis: The Inspired Study."

[c] Cho, Kim, and Chung, "Identification of Novel Population Clusters with Different Susceptibilities to Type 2 Diabetes and Their Impact on the Prediction of Diabetes."

[d] O. P. Zaharia et al., "Risk of Diabetes-Associated Diseases in Subgroups of Patients with Recent-Onset Diabetes: A 5-Year Follow-up Study," *Lancet Diabetes Endocrinol* 7, no. 9 (2019).

e Li et al., "Identification of Type 2 Diabetes Subgroups through Topological Analysis of Patient Similarity."

f Kahkoska et al., "Validation of Distinct Type 2 Diabetes Clusters and Their Association with Diabetes Complications in the Devote, Leader and Sustain-6 Cardiovascular Outcomes Trials."

g Karpati et al., "Patient Clusters Based on HbA1c Trajectories: A Step toward Individualized Medicine in Type 2 Diabetes."

h Arevalo Lorido et al., "Comorbidity in Patients with Type 2 Diabetes Mellitus and Heart Failure with Preserved Ejection Fraction. Cluster Analysis of the Rica Registry. Opportunities for Improvement."

[i] Alexandre et al., "Identification of Diabetes Self-Management Profiles in Adults: A Cluster Analysis Using Selected Self-Reported Outcomes."

[j] Dennis et al., "Disease Progression and Treatment Response in Data-Driven Subgroups of Type 2 Diabetes Compared with Models Based on Simple Clinical Features: An Analysis Using Clinical Trial Data."

[k] Amato et al., "Phenotyping of Type 2 Diabetes Mellitus at Onset on the Basis of Fasting Incretin Tone: Results of a Two-Step Cluster Analysis."

collected. Only participants with a self-reported T2D diagnosis, verified by a nurse, were selected and included in the analyses that follow.

Selection of cardiometabolic markers

The clustering analysis employs 16 common biomarkers of T2D identified through a review of the literature[236,237,238] and review by a clinical domain expert. The 16 biomarkers included are systolic blood pressure, diastolic blood pressure, pulse rate, age, WC, BMI, apolipoprotein A (ApoA), apolipoprotein B (ApoB), CRP, cholesterol, HDL-C, LDL direct, glucose, HbA1c, lipoprotein A, and triglycerides.

Age is an important risk factor for T2D and is included in several models and scores to calculate the risk of diabetes.[239] Age and age at diagnosis are also associated with the risk of complications, with younger age at diagnosis related to a higher risk of mortality and vascular disease.[240] More specifically, studies have found that older age or age at diagnosis and longer duration of diabetes increase the risk of macrovascular events and death, while the negative impact of longer diabetes duration on microvascular event risk diminishes with older age or age at diagnosis, resulting in the highest microvascular risks in the youngest individuals with long-term diabetes.[241] Another study found older individuals with diabetes to have a higher incidence of complications, but overall better glycaemic control,[242] further highlighting the importance of age in differentiating between the phenotypical characteristics of this condition.

BMI and WC are among the top predictors of T2D,[243] and there is evidence of a causal relationship between WC and T2D risk.[244] Values differ among diabetes patients depending on the mechanisms underlying the condition and potential diabetes subgroup.[245] Additionally, higher BMIs are associated with higher risks of adverse outcomes due to diabetes, including insulin dependency, complications of the cardiovascular and renal systems, lower extremity complications, and ocular complications.[246] On the other hand, the Diabetes Remission Clinical Trial (DiRECT)[247] showed that weight loss is a predictor of diabetes remission, with a weight loss greater than 10 kg able to reverse most early T2D cases.[248]

BMI has historically been used to diagnose obesity, with values over 25 kg/m^2 indicating overweight, and over 30 kg/m^2 obesity,[249] and has been a powerful tool in epidemiological research into the obesity pandemic.[250] However, the validity of BMI has been questioned recently, especially in relation to certain demographics, such as post-menopausal women, for whom the conventional cut-off for obesity is not an appropriate indicator of obesity status.[251] Some have even termed BMI as unethical, as the current "healthy" ranges are modelled on the Caucasian population and are not appropriate for other ethnicities.[252] WC is often a preferred marker of obesity and T2D risk, as it has been shown to be a stronger predictor of obesity than BMI.[253]

As diabetes is the clinical manifestation of impaired glucose metabolism, it is not surprising that glucose levels are one of the main biomarkers and predictors of T2D. Impaired fasting glucose is denoted by glucose—measured after at least 8 hours of fasting—values between the normal range of 5.6 mmol/L and the limit for diabetes of 7 mmol/L, and studies have shown that impaired fasting glucose increases the risk of developing T2D.[254] Alternatively, two-hour (2-h) post-glucose-load glucose levels testing is used to diagnose T2D, with values greater than 200 mg/dL after 75 g oral glucose indicating the presence of T2D.[255] The type of test (fasting or 2-h) can give an indication of whether the diabetes is due to insulin resistance or β-cell function loss. Elevated fasting glucose levels may suggest that the body has trouble maintaining normal blood glucose levels overnight and are linked to hepatic insulin resistance.[256] On the other hand, elevated 2-h glucose levels indicate that the body is not effectively using insulin to lower blood glucose levels, which indicates poor β-cell function,[257] as well as insulin resistance in the muscles.[258] Additionally, 2-h glucose has been found to be associated with the risk of CVD, unlike fasting glucose.[259]

Similar to glucose, HbA1c is used in the diagnosis of T2D, with values above 6.5% recommended to diagnose diabetes.[260] HbA1c reflects the average glucose levels over 12 weeks,[261] and it has been found to be the strongest predictor of T2D incidence over a ten-year period.[262] Additionally, consistently high levels of HbA1c are associated with all-cause mortality and cardiac events, and early management of glucose impairment can mitigate these risks.[263] In combination with fasting glucose, HbA1c might be useful to identify the diabetes subtype characterised by β-cell dysfunction rather than insulin resistance. However, it might not be a marker of cardiovascular risk for all people with T2D. For example, for patients with cystic fibrosis-related diabetes, HbA1c underestimates the degree of hyperglycaemia.[264] Together with other markers of T2D, HbA1c is an important factor that can help segment different disease subgroups. HbA1c levels below 42 mmol/mol (or below 6%) are considered normal.

Hypertension and T2D share common risk factors, including lifestyle choices such as physical activity and diet. The two conditions also share several pathophysiological aspects, and therefore, people with T2D are twice as likely to have hypertension compared to people without diabetes. The relationship between hypertension and T2D is bidirectional: not only does high blood pressure increase the risk of developing diabetes, but diabetes also raises the likelihood of developing hypertension due to various pathophysiological mechanisms, including insulin resistance and endothelial dysfunction. Additionally, both hypertension and diabetes are important risk factors for CVD, with the effects of insulin resistance and impaired glucose tolerance on endothelial function being comparable to those caused by hypertension. Petrie et al.[265] provided a comprehensive overview of the interplay between diabetes, hypertension, and CVD. Studies have shown that tight control of blood pressure can significantly decrease the incidence of microvascular and macrovascular complications in people with T2D,[266] as well as being an effective strategy to prevent the onset of T2D.[267] Hypertension is diagnosed when systolic and diastolic blood pressure is above 130 and 80 mmHg, respectively.

Apolipoproteins A and B are both helpful biomarkers for evaluating the risk of T2D, as they are involved in the metabolism of lipoprotein and blood glucose. ApoA is carried in HDL-C, and it aids the removal of detrimental types of cholesterol associated with increased atherosclerosis risk. As such, increased values are associated with a lower risk of coronary artery disease, as well as lower diabetes prevalence.[268] ApoB is involved in the metabolism of lipids and is associated with several diseases, including T2D, stroke, and CVD. Recent studies have shown that ApoB may have a causal association with T2D, as increased ApoB leads to greater uptake of these particles by the pancreas, potentially causing cell dysfunction and reduced insulin secretion, thus contributing to the development of diabetes.[269] For ApoA, the normal range is 110–180 mg/dL for men, while for women, the normal range is 110–205 mg/dL. Normal levels of ApoB-100 in adults are less than 100 mg/dL.

Lipoprotein A (Lp(a)) is an independent and causal risk factor for atherosclerotic CVDs; it is predominantly genetically determined.[270] Lp(a) is most commonly associated with incident CVD, and it has been recently shown to be an independent predictor of recurrent cardiovascular events in people with T2D.[271] Results from observational and epidemiological studies suggest that there is an inverse association between Lp(a) concentrations and T2D, although conclusive evidence of a causal association is lacking.[272] Normal Lp(a) levels are below 30 mg/dL.

Triglycerides significantly increase the risk of atherosclerosis, heart disease, and stroke. Triglyceride levels are a risk factor and predictor of T2D, independent of age, BMI, hypertension, and hyperuricaemia.[273] After the onset of diabetes, the incidence of hypertriglyceridemia is high and associated with significantly higher cardiovascular risk.[274] Hypertriglyceridemia can be improved through lifestyle interventions combining dietary changes, exercise, and reduction of alcohol intake, which can reduce triglyceride levels by up to 60%.[275] Normal triglyceride levels are below 1.7 mmol/L, or 150 mg/dL.

Similar to triglycerides, total cholesterol is a well-known marker for CVD. Patients with T2D are at especially high risk of Hypercholesterolaemia, because insulin resistance and hyperglycaemia result in the overproduction of triglycerides in the liver, as well as decreased clearance of these molecules[276]—these are both associated with increased low-density lipoprotein cholesterol (LDL-C) and decreased HDL-C. While the lipid profile in poorly controlled type 1 diabetes is characterised by low HDL and high LDL, the lipid profile in T2D involves mainly lowered HDL levels, with total cholesterol and LDL only slightly raised. While controlled type 1 diabetes results in a normal lipidic profile, controlling glycaemic metabolism in T2D does not improve lipid levels, highlighting the negative role of insulin resistance on lipid levels.[277] Normal ranges for total cholesterol and LDL-C are below 5 and 4 mmol/L, respectively, while HDL-C should be above 1 mmol/L for men and above 1.2 mmol/L for women.

Chronic inflammation, characterised by elevated levels of inflammatory proteins, plays a significant role in the development of T2D. CRP is a key inflammatory marker produced by liver cells, whose high levels are associated with chronic inflammation and have been linked to risk factors such as obesity, hypertension, heavy alcohol consumption, smoking, and physical inactivity.[278] CRP values less than 0.3 mg/dL are considered within the normal range.

Selection of disease outcomes

Illnesses were self-reported in a questionnaire at the baseline visit and subsequently verified by a nurse through a verbal interview. Three conditions associated with cardiometabolic health were selected: CVD, high cholesterol (or hypercholesterolaemia), and hypertension. The following diseases were included within CVD: heart/cardiac problem, peripheral vascular disease, venous thromboembolic disease, angina, heart attack/myocardial infarction, heart failure/pulmonary oedema, heart arrhythmia, heart valve problem/heart murmur, cardiomyopathy, pericardial problem, stroke, transient ischaemic attack, subdural haemorrhage/haematoma, subarachnoid haemorrhage, leg claudication/intermittent claudication, arterial embolism, pulmonary embolism, deep venous thrombosis, cerebral aneurysm, myocarditis, atrial fibrillation, rheumatic fever, atrial flutter, Wolff–Parkinson–White syndrome, irregular heartbeat, sick sinus syndrome, supraventricular tachycardia, brain haemorrhage, aortic aneurysm, other venous/lymphatic disease, varicose veins, lymphoedema, ischaemic stroke, mitral valve disease, aortic valve disease, hypertrophic cardiomyopathy, pericarditis, pericardial effusion, aortic aneurysm rupture, aortic dissection, varicose ulcer, mitral valve prolapse, mitral stenosis, mitral regurgitation/incompetence, aortic stenosis, and aortic regurgitation/incompetence.

Clustering methodology

Clustering was done on 16 cardiometabolic markers measured at baseline: systolic blood pressure, diastolic blood pressure, pulse rate, age at diagnosis, WC, BMI, ApoA, ApoB, CRP, cholesterol, HDL-C, LDL direct, glucose, HbA1c, lipoprotein A, and triglycerides. All data were rescaled such that variables lie in the [0,1] range. The dataset was split by sex to avoid sex-dependent differences. Four clustering methods were tested: k-means, k-medoids, agglomerative hierarchical clustering, and Gaussian mixture models (GMM). Hyperparameters and the number of clusters were tuned based on the silhouette width.[279] Cluster stability was evaluated using the bootstrapped Jaccard coefficient in the R *fpc* package,[280] which computes the similarity between clusters on the whole sample and clusters on the bootstrapped dataset (1,000 resamples). A cluster with a Jaccard coefficient value ≥0.75 was considered stable.[281] The quality of the clustering methods were evaluated based on the silhouette, Calisnki–Harabasz and Davies–Bouldin scores.

Differences between clusters were assessed with one-way analysis of variance (ANOVAs) for continuous variables and χ^2 tests for categorical variables. Bonferroni correction was applied to the p-value for multiple tests. Effect sizes are reported as Cohen's D for the ANOVA tests and as Cramér's V for the χ^2 tests.

Results

After removing participants with missing values in one or more biomarkers, 13,771 participants were included in the cluster analysis. Descriptive statistics for the clustering dataset are presented in Table 1.2. Additionally, sex-dependent differences in the 16 cardiometabolic markers are shown in Table 1.3.

The four clustering methods, k-means, k-medoids, agglomerative hierarchical clustering, and GMM, were tested with the *scikit-learn*[282] and *scikit-learn-extra* in Python. The clustering methods were evaluated based on the silhouette, Calisnki–Harabasz and Davies–Bouldin scores. GMM clusters with n = 4 clusters and a spherical covariance matrix were chosen. The four clusters generated by GMM achieved a Jaccard score ≥0.75 and therefore were deemed stable.

Cluster 1 is characterised by lower T2D onset age, lower BMI and WC, and improved cardiometabolic profile. Cluster 2 has an average T2D onset age (around 51 years for both males and females), higher BMI, and slightly worse blood pressure and CRP than Cluster 1. Clusters 3 and 4 are characterised by higher onset age and worse cardiometabolic profile, with Cluster 4 having the highest values across all clusters for blood pressure, lipids (total cholesterol, HDL-C, LDL-C), ApoA, ApoB, and Lp(a), glucose, HbA1c, and CRP. Biomarker differences between clusters are presented in Table 1.4 for the male sample and in Table 1.5 for the female sample. A summary of cluster characteristics is shown in Table 1.6. A visual depiction of cluster-specific data is presented in Figure 1.2.

TABLE 1.2
Descriptive table of biomarkers for clustering sample and regression sample

	Clustering sample	UK Biobank
n	13,771	Biomarker-dependent
Males %	64	46
T2D onset age (years)	50.94 (12.64)	N/A
BMI (kg/m²)	31.29 (5.81)	27.33 (4.77)
WC (cm)	102.95 (14.25)	90.17 (13.42)
HbA1c (mmol/mol)	52.93 (13.67)	36.14 (6.75)
Glucose (mmol/L)	7.56 (3.37)	5.12 (1.24)
Diastolic BP (mmHg)	81.37 (10.33)	82.46 (10.76)
Systolic BP (mmHg)	143.32 (18.5)	138.75 (19.35)
Pulse rate (bpm)	74.25 (13.3)	68.87 (11.65)
ApoA (g/L)	1.4 (0.25)	1.54 (0.27)
ApoB (g/L)	0.84 (0.21)	1.03 (0.24)
Lp(a) (nmol/L)	43.64 (49.24)	44.64 (49.21)
Triglycerides (mmol/L)	2.08 (1.17)	1.75 (1.02)
Cholesterol (mmol/L)	4.47 (0.99)	5.69 (1.15)
LDL direct (mmol/L)	1.19 (0.32)	3.56 (0.87)
HDL-C (mmol/L)	2.68 (0.74)	1.45 (0.38)
CRP	3.47 (5.2)	2.59 (4.35)

Data presented as means and standard deviation.

TABLE 1.3
Sex-dependent differences in cardiometabolic biomarkers

Biomarker	Males (n = 8,929)	Females (n = 5,655)	p-value of difference*	Standardised variance
Age (years)	60.02 (6.98)	58.89 (7.48)	<0.001	0.048
BMI (kg/m²)	30.8 (5.31)	31.86 (6.52)	<0.001	0.017
WC (cm)	105.3 (13.38)	97.76 (14.91)	<0.001	0.018
HbA1c (mmol/mol)	52.61 (13.33)	51.73 (13.53)	<0.001	0.018
Glucose (mmol/L)	7.57 (3.3)	7.21 (3.15)	<0.001	0.021
Systolic BP (mmHg)	143.73 (18.25)	142.04 (19.08)	<0.001	0.013
Diastolic BP (mmHg)	81.7 (10.25)	80.82 (10.44)	<0.001	0.013
Pulse rate (bpm)	73.17 (13.47)	75.64 (12.53)	<0.001	0.016
ApoA (g/L)	1.34 (0.22)	1.51 (0.25)	<0.001	0.012
ApoB (g/L)	0.83 (0.2)	0.88 (0.23)	<0.001	0.021
Lp(a) (nmol/L)	43.11 (49.01)	44.7 (49.58)	0.057	0.07
Triglycerides (mmol/L)	2.1 (1.15)	1.97 (1.05)	<0.001	0.022
Cholesterol (mmol/L)	4.31 (0.95)	4.8 (1.04)	<0.001	0.019
LDL direct (mmol/L)	2.61 (0.71)	2.86 (0.8)	<0.001	0.016
HDL-cholesterol (mmol/L)	1.12 (0.28)	1.34 (0.34)	<0.001	0.017
CRP	2.68 (3.31)	4.03 (4.68)	<0.001	0.015

TABLE 1.4
Cardiometabolic markers in the clustering dataset for whole samples and each cluster, for males

Variable	Whole sample (n = 8,870)	Cluster 1 (n = 1,585)	Cluster 2 (n = 1,347)	Cluster 3 (n = 1181)	Cluster 4 (n = 788)
WC	98.7 (14.7)	92.4 (13.5)*‡§	100.4 (15.4)†§	101.8 (13.5)†§	103.8 (12.8)†*‡
PR	76.1 (12.7)	74.3 (12.9)*‡§	75.8 (12.1)†§	77.1 (12.9)†	78.2 (13.1)†*
Age	51.1 (12.5)	47.4 (10.8)*‡§	52.8 (14.9)†	53.2 (10.8)†	52.8 (11)†
D-BP	80.8 (10.4)	79 (10.2)*‡§	81.6 (10.5)†	81.5 (10.2)†	81.5 (10.5)†
S-BP	142.6 (18.9)	140.4 (18.4)*‡§	143.2 (18.5)†	143.5 (18.4)†	144.6 (18.9)†
BMI	32.2 (6.5)	30.07 (6.3)*‡§	33 (6.6)†	33.3 (6.3)†	33.4 (5.9)†
ApoA	1.5 (0.3)	1.6 (0.2)*‡§	1.5 (0.3)†‡§	1.5 (0.2)†*§	1.4 (0.2)†*‡
ApoB	0.9 (0.2)	0.8 (0.2)*‡§	0.9 (0.2)†‡§	0.9 (0.2)†*§	1 (0.2)†*‡
Chol	4.7 (1)	4.5 (0.9)*‡§	4.6 (0.9)†‡§	4.8 (0.9)†*§	5.3 (1.1)†*‡
CRP	4.3 (5.7)	3.4 (5.9)*‡§	4.59 (5.4)†	4.7 (5.9)†	5.1 (5.5)†
Glucose	7.43 (3.34)	7.2 (3)§	7.1 (3.4)‡§	7.6 (3)*§	8.1 (3.7)†*‡
HbA1c	53.1 (13.6)	52.1 (12.7)*‡§	51.8 (13)‡§	54.2 (12.7)†*	55.6 (14.3)†*
HDL-C	1.3 (0.3)	1.5 (0.3)*‡§	1.3 (0.4)†‡§	1.2 (0.3)†*§	1.1 (0.2)†*‡
LDL-D	2.8 (0.8)	2.5 (0.7)*‡§	2.8 (0.7)†‡§	2.9 (0.7)†*§	3.2 (0.8)†*‡
Lp(a)	44.4 (49.5)	49.2 (50)‡§	44.7 (51.3)	41 (50)†	39 (46.7)†
TG	2 (1.1)	1 (0.2)*‡§	1.7 (0.3)†‡§	2.4 (0.2)†*§	4 (1.1)†*‡

WC: waist circumference; PR: pulse rate; D-BP: diastolic blood pressure; S-BP: systolic blood pressure; BMI: body mass index; ApoA: apolipoprotein A, ApoB: apolipoprotein B; Chol: cholesterol; CRP: C-reactive protein; HbA1c: glycated haemoglobin 1C; HDL-C: high density lipoprotein cholesterol; LDL-C: low density lipoprotein cholesterol; Lp(a): lipoprotein A; TG: triglycerides.

† Difference significant with Cluster 1, * difference significant with Cluster 2, ‡ difference significant with Cluster 3, § difference significant with Cluster 4.

The differences between clusters are similar in males and females, as shown in Table 1.4 and Table 1.5. However, the prevalence of comorbidities differs between clusters, with males in Cluster 3 having the highest prevalence of hypertension and CVD (31% and 65%, respectively) and males in Cluster 4 having the highest prevalence of Hypercholesterolaemia (40%). For females, Cluster 4 has the highest prevalence of Hypercholesterolaemia (43%) and CVD (25%), while Cluster 3 has the highest prevalence of hypertension (67%). Cluster 1, for both males and females, was characterised by significantly higher insulin use (28% in males, 33% in females vs <19% in other clusters).

Discussion

Four phenotypically distinct clusters of T2D were found based on 16 cardiometabolic markers routinely measured in clinical settings. The clustering highlighted the presence of different types of T2D, characterised by different cardiometabolic profiles. For example, one cluster (Cluster 1) presented low BMI and lower onset age of

TABLE 1.5
Cardiometabolic markers in the clustering dataset for whole samples and each cluster, for females

Variable	Whole sample (n = 4,901)	Cluster 1 (n = 2,643)	Cluster 2 (n = 2,366)	Cluster 3 (n = 2,183)	Cluster 4 (n = 1,678)
WC	105.3 (13.4)	100.6 (12.6)*‡§	105.7 (13.2)†‡§	107.8 (13.5)†*	108.9 (12.6)†*
PR	73.3 (13.5)	71.2 (13.2)*‡§	72.3 (13.4)†‡§	74.3 (13.4)†*§	76.5 (13.6)†*‡
Age	50.8 (12.7)	48.1 (11.2)*‡§	51.8 (15.1)†	52.1 (12.2)†	52.1 (10.4)†
D-BP	81.7 (10.3)	80.1 (10)*‡§	81.8 (10.6)†‡§	82.2 (10.1)†§	83.4 (9.9)†*‡
S-BP	143.7 (18.3)	142.3 (18.7)‡§	143.7 (18.2)§	144.2 (18.3)†	145.3 (17.6)†*
BMI	30.8 (5.3)	29.1 (5.1)*‡§	30.9 (5.1)†‡§	31.8 (5.46)†*	32.1 (5)†*
ApoA	1.3 (0.2)	1.4 (0.2)*‡§	1.3 (0.25)†‡§	1.3 (0.2)†*§	1.3 (0.2)†*‡
ApoB	0.8 (0.2)	0.7 (0.2)*‡§	0.8 (0.17)†‡§	0.9 (0.2)†*§	0.9 (0.2)†*‡
Chol	4.3 (1)	4 (0.9)*‡§	4.2 (0.8)†‡§	4.4 (0.9)†*§	4.8 (1.1)†*‡
CRP	3 (4.9)	2.7 (4.7)‡§	3 (5.2)†	3.1 (4.9)†	3.3 (4.3)†
Glucose	7.6 (3.4)	7.3 (3.)‡§	7.34 (3.27)‡§	7.7 (3.2)†*§	8.5 (3.8)†*‡
HbA1c	52.8 (13.7)	51.6 (13.5)‡§	51.7 (13.2)‡§	53.4 (13)†*§	55.7 (15.2)†*‡
HDL-C	1.1 (0.3)	1.27 (0.2)*‡§	1.1 (0.3)†‡§	1.1 (0.2)†*§	1 (0.2)†*‡
LDL-D	2.6 (0.7)	2.36 (0.7)*‡§	2.57 (0.6)†‡§	2.7 (0.7)†*§	2.9 (0.8)†*‡
Lp(a)	43.2 (49.1)	47.6 (47.4)*‡§	42.7 (51.2)†	41 (48.4)†	40.1 (48.5)†
TG	2.1 (1.2)	1 (0.3)*‡§	1.7 (0.2)†‡§	2.4 (0.2)†*§	4.1 (1.17)†*‡

† Difference significant with Cluster 1, * difference significant with Cluster 2, ‡ difference significant with Cluster 3, § difference significant with Cluster 4.

WC: waist circumference; PR: pulse rate; D-BP: diastolic blood pressure; S-BP: systolic blood pressure; BMI: body mass index; AL-A: apolipoprotein A, AL-B: apolipoprotein B; Chol: cholesterol; CRP: C-reactive protein; HbA1c: glycated haemoglobin 1C; HDL-C: high density lipoprotein cholesterol; LDL-C: low density lipoprotein cholesterol; Lp(a): lipoprotein A; TG: triglycerides.

diabetes, while maintaining better levels of triglycerides and lipid levels compared to the other clusters. This, together with the increased insulin use in the cluster, suggests that this kind of diabetes subtype may not be caused by an unhealthy lifestyle but rather by impaired mechanisms of glucose control. On the other hand, Clusters 3 and 4 presented higher onset age, obesity, and higher levels of triglycerides and lipids. Paired with the higher incidence of CVD, hypertension, and Hypercholesterolaemia, the clusters may be representative of a diabetes subtype linked to unhealthy diets and inactivity.

The clusters found in this analysis resemble the clusters found by Ahlqvist et al.[283] The authors found one cluster which was insulin-dependent, with low BMI and poor metabolic control (similar to Cluster 1); and two other two clusters with higher onset age, no insulin resistance, and with obesity (similar to Clusters 3 and 4). However, compared to the clusters found by Ahlqvist et al., the clusters described here did not present differences in glucose metabolism (especially HbA1c), but rather highlighted

TABLE 1.6
Summary of cluster characteristics

	Males				Females			
Variable	C1	C2	C3	C4	C1	C2	C3	C4
Onset age	A	A	A	A	L	A	A	A
BMI/WC	B	A*	A*	W*	B*	A*	A*	W*
BP	A*	A*	A*	A*	A	A*	A*	A*
Cholesterol	B	B	W*	W*	B	A*	A*	W*
Triglycerides	B	B	W*	W*	B	B	W*	W*
Glucose	A*	A*	A*	A*	A*	A*	A*	A*
HbA1c	A*	A*	A*	A*	A*	A*	A*	A*
CRP	A*	A*	A*	A*	A*	A*	A*	A*
Highest odds	Insulin use		HT, CVD	HC	Insulin use		HT	HC, CVD

Biomarker values beyond the normal range are marked with "*". CVD: cardiovascular disease, HT: hypertension, HC: high cholesterol.

Biomarkers are marked as "better" (B) or "worse" (W) if the Cohen's d in t-tests between the cluster and the whole sample (stratified by sex) was ≥0.25, otherwise they are marked as "average" (A)

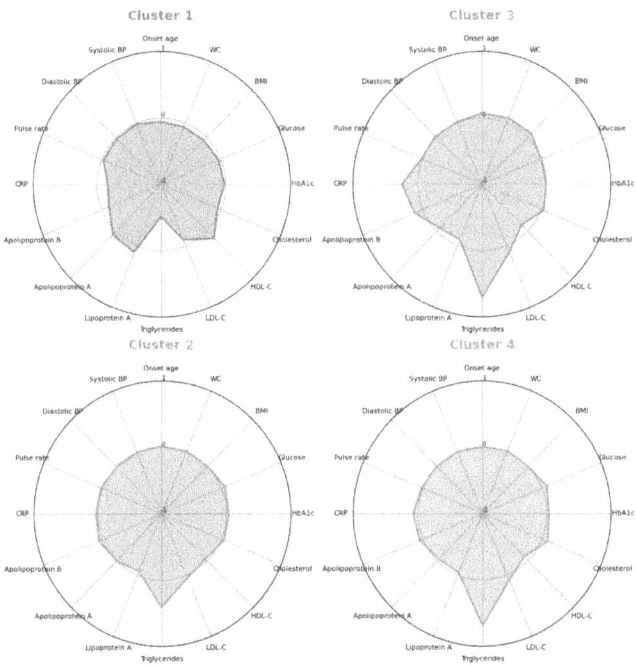

FIGURE 1.2 Visual representation of biomarkers in the four clusters. The data are presented as the normalised difference between the cluster average and the sample average. Data averaged across males and females.

the differences in overall cardiometabolic health, suggesting that the differences in diabetes clusters could be a product of lifestyle differences. The case study in Chapter 2 will explore the differences in physical behaviours between the clusters.

Part of this case study has been published in "Personalised lifestyle recommendations for type 2 diabetes: Design and simulation of a recommender system on UK Biobank Data." Cavallo F. R., Toumazou C. (2023) *PLOS Digital Health* 2(8): e0000333. https://doi.org/10.1371/journal.pdig.0000333

ABBREVIATIONS

AI	Artificial intelligence
ANOVA	Analysis of variance
ApoA	Apolipoprotein A
ApoB	Apolipoprotein B
BMI	Body mass index
CDSS	Clinical decision support system
CRP	C-reactive protein
CVD	Cardiovascular disease
FTO	Fat mass and obesity-associated
GADA	Glutamic acid decarboxylase antibodies
GIP	Glucose-dependent insulinotropic polypeptide
GLP-1	Glucagon-like peptide-1
GMM	Gaussian mixture models
GRS	Genetic risk score
GWAS	Genome-wide association studies
HbA1c	Glycated haemoglobin
HDL-C	High density lipoprotein cholesterol
HOMA-IR	Homeostatic Model Assessment for Insulin Resistance
LDL-C	Low density lipoprotein cholesterol
Lp(a)	Lipoprotein A
MR	Mendelian randomisation
NCD	Non-communicable diseases
PPGR	Post-prandial glucose response
PRS	Polygenic risk score
RCT	Randomised controlled trial
SNP	Single nucleotide polymorphism
T2D	Type 2 diabetes
WC	Waist circumference
WHO	World Health Organization

NOTES

1. World Health Organization, "Non-Communicable Diseases: Country Profiles 2018," World Health Organization, (2018), https://apps.who.int/iris/handle/10665/274512 (Accessed 24/03/2024).

2. GBD Risk Factors Collaborators, "Global, Regional, and National Comparative Risk Assessment of 79 Behavioural, Environmental and Occupational, and Metabolic Risks or Clusters of Risks, 1990–2015: A Systematic Analysis for the Global Burden of Disease Study 2015," *Lancet* 388, no. 10053 (2016): 1659–724.
3. D. Nunan, D. N. Blane, and M. McCartney, "Exemplary Medical Care or Trojan Horse? An Analysis of the 'Lifestyle Medicine' Movement," *Br J Gen Pract* 71, no. 706 (2021): 229–32.
4. S. Chatterjee, K. Khunti, and M. J. Davies, "Type 2 Diabetes," *Lancet* 389, no. 10085 (2017): 2239–51.
5. Z. S. Morris, S. Wooding, and J. Grant, "The Answer Is 17 Years, What Is the Question: Understanding Time Lags in Translational Research," *J R Soc Med* 104, no. 12 (2011): 510–20.
6. Z. Stark et al., "Integrating Genomics into Healthcare: A Global Responsibility," *Am J Hum Genet* 104, no. 1 (2019): 13–20.
7. G. Lewis et al., "The Clinical Effectiveness of Sertraline in Primary Care and the Role of Depression Severity and Duration (Panda): A Pragmatic, Double-Blind, Placebo-Controlled Randomised Trial," *Lancet Psychiatry* 6, no. 11 (2019): 903–14.
8. I. Zucker and B. J. Prendergast, "Sex Differences in Pharmacokinetics Predict Adverse Drug Reactions in Women," *Biol Sex Differ* 11, no. 1 (2020): 32.
9. Council of the European Union, "Council Conclusions on Personalised Medicine for Patients," no. 421 (2015): 2–5.
10. A. D. Weston and L. Hood, "Systems Biology, Proteomics, and the Future of Health Care: Toward Predictive, Preventative, and Personalized Medicine," *J Proteome Res* 3, no. 2 (2004): 179–96.
11. L. Hood, "A Systems Approach to Medicine Will Transform Healthcare," in *Physical Biology*, 337–66 (Published by Imperial College Press and distributed by World Scientific Publishing Co., 2008).
12. L. Hood and C. Auffray, "Participatory Medicine: A Driving Force for Revolutionizing Healthcare," *Genome Med* 5, no. 12 (2013): 110.
13. G. P. Patrinos et al., "Using Chatgpt to Predict the Future of Personalized Medicine," *Pharmacogenomics J* 23, no. 6 (2023): 178–84.
14. W. E. Evans and J. A. Johnson, "Pharmacogenomics: The Inherited Basis for Interindividual Differences in Drug Response," *Annu Rev Genomics Hum Genet* 2, no. 1 (2001): 9–39.
15. M. Whirl-Carrillo, T. E. Klein, and R. B. Altman, "Pharmacogenomics," in *Brenner's Encyclopedia of Genetics*, ed. Stanley Maloy and Kelly Hughes, 283–85 (San Diego: Academic Press, 2013).
16. National Cancer Institute, "Precision Medicine," (2024), https://www.cancer.gov/publications/dictionaries/cancer-terms/def/precision-medicine (Accessed 29/09/2024).
17. International Human Genome Sequencing Consortium, "Finishing the Euchromatic Sequence of the Human Genome," *Nature* 431, no. 7011 (2004): 931–45.
18. The International HapMap Consortium, "A Haplotype Map of the Human Genome," *Nature* 437, no. 7063 (2005): 1299–320.
19. Encode Project Consortium et al., "Identification and Analysis of Functional Elements in 1% of the Human Genome by the Encode Pilot Project," *Nature* 447, no. 7146 (2007): 799–816.
20. The Lancet Digital Health, "Digital Technologies: A New Determinant of Health," *Lancet Digit Health* 3, no. 11 (2021): e684.
21. S. Schleidgen et al., "What Is Personalized Medicine: Sharpening a Vague Term Based on a Systematic Literature Review," *BMC Med Ethics* 14, no. 1 (2013): 55.
22. G. De Grandis and V. Halgunset, "Conceptual and Terminological Confusion around Personalised Medicine: A Coping Strategy," *BMC Med Ethics* 17 (2016): 43.

23. H. Doctor-Loeb, "Precision Medicine Has Been Overhyped," (2023), https://www.scientificamerican.com/article/precision-medicine-has-been-overhyped/ (Accessed 28/02/2024).
24. R. Bayer and S. Galea, "Public Health in the Precision-Medicine Era," *N Engl J Med* 373, no. 6 (2015): 499–501.
25. S. Senn, "Statistical Pitfalls of Personalised Medicine," *Nature* 563 (2018): 619–21.
26. A. V. Khera et al., "Genome-Wide Polygenic Scores for Common Diseases Identify Individuals with Risk Equivalent to Monogenic Mutations," *Nat Genet* 50, no. 9 (2018): 1219–24.
27. S. Cuschieri, "The Genetic Side of Type 2 Diabetes – A Review," *Diabetes Metab Syndr* 13, no. 4 (2019): 2503–6.
28. N. J. Schork, "Personalized Medicine: Time for One-Person Trials," *Nature* 520, no. 7549 (2015): 609–11.
29. J. Nikles et al., "Establishment of an International Collaborative Network for N-of-1 Trials and Single-Case Designs," *Contemp Clin Trials Commun* 23 (2021): 100826.
30. Senn, "Statistical Pitfalls of Personalised Medicine."
31. J. P. Samuel, S. H. Wootton, and J. E. Tyson, "N-of-1 Trials: The Epitome of Personalized Medicine?," *J Clin Transl Sci* 7, no. 1 (2023): e161.
32. J. Wilkinson et al., "Time to Reality Check the Promises of Machine Learning-Powered Precision Medicine," *Lancet Digit Health* 2, no. 12 (2020): e677-e80.
33. Hippocrates, "Hippocratic Writings," ed. G. E. R. Lloyd and Chadwick J. (Harmondsworth: Penguin Classics, 2015).
34. F. Lagay, "The Legacy of Humoral Medicine," *Virtual Mentor* 4, no. 7 (2002): 206–8.
35. R. D. Mirza et al., "The History and Development of N-of-1 Trials," *J R Soc Med* 110, no. 8 (2017): 330–40.
36. Lagay, "The Legacy of Humoral Medicine."
37. Science Museum, "The Microscope," (2019), https://www.sciencemuseum.org.uk/objects-and-stories/medicine/microscope#:~:text=It's%20not%20clear%20who%20invented,30%20times%20its%20normal%20size (Accessed 18/02/2024).
38. Mark Mosley, "Viewpoint: Germ Theory: A History of Truth and Denial," *Emerg Med News* 42, no. 9C (2020).
39. International Human Genome Sequencing Consortium, "Finishing the Euchromatic Sequence of the Human Genome."
40. Hood, "A Systems Approach to Medicine Will Transform Healthcare."
41. M. Vailati-Riboni, V. Palombo, and J. J. Loor, "What Are Omics Sciences?," in *Periparturient Diseases of Dairy Cows*, ed. Burim N. Ametaj, 1–7 (Cham: Springer International Publishing, 2017).
42. Vailati-Riboni, Palombo, and Loor, "What Are Omics Sciences?"
43. K. V. Voelkerding, S. A. Dames, and J. D. Durtschi, "Next-Generation Sequencing: From Basic Research to Diagnostics," *Clin Chem* 55, no. 4 (2009): 641–58.
44. R. Aebersold and M. Mann, "Mass Spectrometry-Based Proteomics," *Nature* 422, no. 6928 (2003): 198–207.
45. Aebersold and Mann, "Mass Spectrometry-Based Proteomics."
46. J. Jumper et al., "Highly Accurate Protein Structure Prediction with Alphafold," *Nature* 596, no. 7873 (2021): 583–89.
47. DeepMind and EMBL-EBI, "Alphafold Protein Structure Database," https://alphafold.ebi.ac.uk/about (Accessed 22/02/2024).
48. Polygenic risk scores are scores that indicate the susceptibility of an individual to disease based on their genetic traits and the importance of those traits for a specific condition. The importance of genetic traits corresponds to the effect sizes obtained from GWAS.

49. P. M. Visscher et al., "10 Years of Gwas Discovery: Biology, Function, and Translation," *Am J Hum Genet* 101, no. 1 (2017): 5–22.
50. Visscher et al., "10 Years of Gwas Discovery: Biology, Function, and Translation."
51. M. Inouye et al., "Genomic Risk Prediction of Coronary Artery Disease in 480,000 Adults: Implications for Primary Prevention," *J Am Coll Cardiol* 72, no. 16 (2018): 1883–93.
52. E. Sanderson et al., "Mendelian Randomization," *Nat Rev Methods Primers* 2, no. 1 (2022): 605–13.
53. A. Doherty et al., "Gwas Identifies 14 Loci for Device-Measured Physical Activity and Sleep Duration," *Nat Commun* 9, no. 1 (2018): 5257.
54. K. J. A. Johnston et al., "Genome-Wide Association Study of Multisite Chronic Pain in UK Biobank," *PLoS Genet* 15, no. 6 (2019): e1008164.
55. D. M. Howard et al., "Genome-Wide Association Study of Depression Phenotypes in UK Biobank Identifies Variants in Excitatory Synaptic Pathways," *Nat Commun* 9, no. 1 (2018): 1470.
56. Doherty et al., "Gwas Identifies 14 Loci for Device-Measured Physical Activity and Sleep Duration."
57. E. Zengini et al., "Genome-Wide Analyses Using UK Biobank Data." Provide Insights into the Genetic Architecture of Osteoarthritis," *Nat Genet* 50, no. 4 (2018): 549–58.
58. S. R. Frank, "Digital Health Care – The Convergence of Health Care and the Internet," *J Ambul Care Manage* 23, no. 2 (2000): 8–17.
59. C. Guo et al., "Challenges for the Evaluation of Digital Health Solutions – A Call for Innovative Evidence Generation Approaches," *NPJ Digit Med* 3, no. 1 (2020): 110.
60. A. Haleem et al., "Telemedicine for Healthcare: Capabilities, Features, Barriers, and Applications," *Sens Int* 2 (2021): 100117.
61. R. T. Sutton et al., "An Overview of Clinical Decision Support Systems: Benefits, Risks, and Strategies for Success," *NPJ Digit Med* 3, no. 1 (2020): 17.
62. R. S. Evans, "Electronic Health Records: Then, Now, and in the Future," *Yearb Med Inform* Suppl 1, no. Suppl 1 (2016): S48–61.
63. Sutton et al., "An Overview of Clinical Decision Support Systems: Benefits, Risks, and Strategies for Success."
64. Y. Li et al., "Impact of Healthy Lifestyle Factors on Life Expectancies in the US Population," *Circulation* 138, no. 4 (2018): 345–55.
65. V. Miller et al., "Evaluation of the Quality of Evidence of the Association of Foods and Nutrients with Cardiovascular Disease and Diabetes: A Systematic Review," *JAMA Netw Open* 5, no. 2 (2022): e2146705.
66. P. Posadzki et al., "Exercise/Physical Activity and Health Outcomes: An Overview of Cochrane Systematic Reviews," *BMC Public Health* 20, no. 1 (2020): 1724.
67. T. J. Saunders et al., "Sedentary Behaviour and Health in Adults: An Overview of Systematic Reviews," *Appl Physiol Nutr Metab* 45, no. 10 (Suppl. 2) (2020): S197-S217.
68. M. J. Franz et al., "Academy of Nutrition and Dietetics Nutrition Practice Guideline for Type 1 and Type 2 Diabetes in Adults: Systematic Review of Evidence for Medical Nutrition Therapy Effectiveness and Recommendations for Integration into the Nutrition Care Process," *J Acad Nutr Diet* 117, no. 10 (2017): 1659–79.
69. A. B. Evert et al., "Nutrition Therapy for Adults with Diabetes or Prediabetes: A Consensus Report," *Diabetes Care* 42, no. 5 (2019): 731–54.
70. J. M. Ordovas et al., "Personalised Nutrition and Health," *BMJ* 361 (2018): bmj k2173.
71. Evert et al., "Nutrition Therapy for Adults with Diabetes or Prediabetes: A Consensus Report."
72. A. J. Scheen, "Precision Medicine: The Future in Diabetes Care?" *Diabetes Res Clin Pract* 117 (2016): 12–21.

73. Ordovas et al., "Personalised Nutrition and Health."
74. Y. C. Klimentidis et al., "Phenotypic and Genetic Characterization of Lower LDL Cholesterol and Increased Type 2 Diabetes Risk in the UK Biobank," *Diabetes* 69, no. 10 (2020): 2194–205.
75. S. F. W. Meddens et al., "Genomic Analysis of Diet Composition Finds Novel Loci and Associations with Health and Lifestyle," *Mol Psychiatry* 26, no. 6 (2021): 2056–69.
76. J. B. Cole, J. C. Florez, and J. N. Hirschhorn, "Comprehensive Genomic Analysis of Dietary Habits in UK Biobank Identifies Hundreds of Genetic Associations," *Nat Commun* 11, no. 1 (2020): 1467.
77. L. Aronica et al., "Genetic Variants for Personalised Management of Very Low Carbohydrate Ketogenic Diets," *BMJ Nutr Prev Health* 3, no. 2 (2020): 363–73.
78. G. Ruano et al., "Physiogenomic Analysis of Weight Loss Induced by Dietary Carbohydrate Restriction," *Nutr Metab (Lond)* 3, no. 1 (2006): 20.
79. Ruano et al., "Physiogenomic Analysis of Weight Loss Induced by Dietary Carbohydrate Restriction."
80. R. Blanco-Rojo et al., "Interaction of an S100a9 Gene Variant with Saturated Fat and Carbohydrates to Modulate Insulin Resistance in 3 Populations of Different Ancestries," *Am J Clin Nutr* 104, no. 2 (2016): 508–17.
81. R. Kang et al., "Consumption of Whole Grains and Legumes Modulates the Genetic Effect of the Apoa5 -1131c Variant on Changes in Triglyceride and Apolipoprotein a-V Concentrations in Patients with Impaired Fasting Glucose or Newly Diagnosed Type 2 Diabetes," *Trials* 15, no. 1 (2014): 100.
82. M. F. Hjorth et al., "Pretreatment Fasting Plasma Glucose and Insulin Modify Dietary Weight Loss Success: Results from 3 Randomized Clinical Trials," *Am J Clin Nutr* 106, no. 2 (2017): 499–505.
83. H. S. Dashti et al., "Gene–Environment Interactions of Circadian-Related Genes for Cardiometabolic Traits," *Diabetes Care* 38, no. 8 (2015): 1456–66.
84. J. S. Zheng et al., "Genetic Variants at Psmd3 Interact with Dietary Fat and Carbohydrate to Modulate Insulin Resistance," *J Nutr* 143, no. 3 (2013): 354–61.
85. T. Huang et al., "PCSK7 Genotype Modifies Effect of a Weight-Loss Diet on 2-Year Changes of Insulin Resistance: The Pounds Lost Trial," *Diabetes Care* 38, no. 3 (2015): 354–61.
86. J. S. Zheng et al., "Modulation by Dietary Fat and Carbohydrate of IRS1 Association with Type 2 Diabetes Traits in Two Populations of Different Ancestries," *Diabetes Care* 36, no. 9 (2013): 2621–7.
87. F. Dudbridge, "Power and Predictive Accuracy of Polygenic Risk Scores," *PLoS Genet* 9, no. 3 (2013): e1003348.
88. J. Yang et al., "Common SNPs Explain a Large Proportion of the Heritability for Human Height," *Nat Genet* 42, no. 7 (2010): 565–9.
89. T. Huang et al., "Genetic Susceptibility to Diabetes and Long-Term Improvement of Insulin Resistance and Beta Cell Function During Weight Loss: The Preventing Overweight Using Novel Dietary Strategies (Pounds Lost) Trial," *Am J Clin Nutr* 104, no. 1 (2016): 198–204.
90. R. Villegas et al., "Gene-Carbohydrate and Gene-Fiber Interactions and Type 2 Diabetes in Diverse Populations from the National Health and Nutrition Examination Surveys (Nhanes) as Part of the Epidemiologic Architecture for Genes Linked to Environment (Eagle) Study," *BMC Genet* 15, no. 1 (2014): 69.
91. J. M. Ordovas et al., "Dietary Fat Intake Determines the Effect of a Common Polymorphism in the Hepatic Lipase Gene Promoter on High-Density Lipoprotein Metabolism: Evidence of a Strong Dose Effect in This Gene-Nutrient Interaction in the Framingham Study," *Circulation* 106, no. 18 (2002): 2315–21.

92. S. Jansen et al., “Effect of 347-Serine Mutation in Apoprotein a-Iv on Plasma Ldl Cholesterol Response to Dietary Fat,” *Arterioscler Thromb Vasc Biol* 17, no. 8 (1997): 1532–8.
93. D. Corella et al., “Alcohol Drinking Determines the Effect of the Apoe Locus on LDL-Cholesterol Concentrations in Men: The Framingham Offspring Study,” *Am J Clin Nutr* 73, no. 4 (2001): 736–45.
94. J. A. Nettleton et al., “Carbohydrate Intake Modifies Associations between Angptl4[E40k] Genotype and HDL-Cholesterol Concentrations in White Men from the Atherosclerosis Risk in Communities (ARIC) Study,” *Atherosclerosis* 203, no. 1 (2009): 214–20.
95. F. Fumeron et al., “Alcohol Intake Modulates the Effect of a Polymorphism of the Cholesteryl Ester Transfer Protein Gene on Plasma High Density Lipoprotein and the Risk of Myocardial Infarction,” *J Clin Invest* 96, no. 3 (1995): 1664–71.
96. M. C. Cornelis et al., “Coffee, CYP1A2 Genotype, and Risk of Myocardial Infarction,” *JAMA* 295, no. 10 (2006): 1135–41.
97. Y. Yang et al., “Effect of Apolipoprotein E Genotype and Saturated Fat Intake on Plasma Lipids and Myocardial Infarction in the Central Valley of Costa Rica,” *Hum Biol* 79, no. 6 (2007): 637–47.
98. E. A. Ruiz-Narvaez, P. Kraft, and H. Campos, “Ala12 Variant of the Peroxisome Proliferator-Activated Receptor-Gamma Gene (PPARG) Is Associated with Higher Polyunsaturated Fat in Adipose Tissue and Attenuates the Protective Effect of Polyunsaturated Fat Intake on the Risk of Myocardial Infarction,” *Am J Clin Nutr* 86, no. 4 (2007): 1238–42.
99. M. C. Cornelis, A. El-Sohemy, and H. Campos, “GSTT Genotype Modifies the Association between Cruciferous Vegetable Intake and the Risk of Myocardial Infarction,” *Am J Clin Nutr*, no. 3 (2007): 752–8.
100. H. Allayee et al., “Nutrigenetic Association of the 5-Lipoxygenase Gene with Myocardial Infarction,” *Am J Clin Nutr* 88, no. 4 (2008): 934–40.
101. S. Hellstrand et al., “Genetic Variation in FADS1 Has Little Effect on the Association between Dietary Pufa Intake and Cardiovascular Disease,” *J Nutr* 144, no. 9 (2014): 1356–63.
102. G. Hindy et al., “The Chromosome 9p21 Variant Interacts with Vegetable and Wine Intake to Influence the Risk of Cardiovascular Disease: A Population-Based Cohort Study,” *BMC Med Genet* 15, no. 1 (2014): 1220.
103. J. Gustavsson et al., “FTO Gene Variation, Macronutrient Intake and Coronary Heart Disease Risk: A Gene-Diet Interaction Analysis,” *Eur J Nutr* 55, no. 1 (2016): 247–55.
104. Y. Zheng et al., “Sugar-Sweetened Beverage Intake, Chromosome 9p21 Variants, and Risk of Myocardial Infarction in Hispanics,” *Am J Clin Nutr* 103, no. 4 (2016): 1179–84.
105. I. Arkadianos et al., “Improved Weight Management Using Genetic Information to Personalize a Calorie Controlled Diet,” *Nutr J* 6, no. 29 (2007): 29.
106. M. Vranceanu et al., “A Comparison of a Ketogenic Diet with a Lowgi/Nutrigenetic Diet over 6 Months for Weight Loss and 18-Month Follow-Up,” *BMC Nutr* 6 (2020): 53.
107. M. Karvela et al., “Assessment of the Impact of a Personalised Nutrition Intervention in Impaired Glucose Regulation over 26 Weeks: A Randomised Controlled Trial,” *Sci Rep* 14, no. 1 (2024): 5428.
108. A. A. Kolodziejczyk, D. Zheng, and E. Elinav, “Diet-Microbiota Interactions and Personalized Nutrition,” *Nat Rev Microbiol* 17, no. 12 (2019): 742–753.
109. P. Kovatcheva-Datchary et al., “Dietary Fiber-Induced Improvement in Glucose Metabolism Is Associated with Increased Abundance of Prevotella,” *Cell Metab* 22, no. 6 (2015): 971–82.

110. D. Zeevi et al., "Personalized Nutrition by Prediction of Glycemic Responses," *Cell* 163, no. 5 (2015): 1079–94.
111. M. Rein et al., "Effects of Personalized Diets by Prediction of Glycemic Responses on Glycemic Control and Metabolic Health in Newly Diagnosed T2dm: A Randomized Dietary Intervention Pilot Trial," *BMC Med* 20, no. 1 (2022): 56.
112. D. Corella et al., "Utilizing Nutritional Genomics to Tailor Diets for the Prevention of Cardiovascular Disease: A Guide for Upcoming Studies and Implementations," *Expert Rev Mol Diagn* 17, no. 5 (2017): 495–513.
113. Kolodziejczyk, Zheng, and Elinav, "Diet-Microbiota Interactions and Personalized Nutrition."
114. S. E. Berry et al., "Human Postprandial Responses to Food and Potential for Precision Nutrition," *Nat Med* 26, no. 6 (2020): 964–73.
115. Corella et al., "Utilizing Nutritional Genomics to Tailor Diets for the Prevention of Cardiovascular Disease: A Guide for Upcoming Studies and Implementations."
116. K. A. Grimaldi et al., "Proposed Guidelines to Evaluate Scientific Validity and Evidence for Genotype-Based Dietary Advice," *Genes Nutr* 12, no. 1 (2017): 35.
117. C. Celis-Morales et al., "Effect of Personalized Nutrition on Health-Related Behaviour Change: Evidence from the Food4me European Randomized Controlled Trial," *Int J Epidemiol* 46, no. 2 (2017): 578–88.
118. C. Celis-Morales et al., "Can Genetic-Based Advice Help You Lose Weight? Findings from the Food4me European Randomized Controlled Trial," *Am J Clin Nutr* 105, no. 5 (2017): 1204–13.
119. Celis-Morales et al., "Can Genetic-Based Advice Help You Lose Weight? Findings from the Food4me European Randomized Controlled Trial."
120. J. R. Horne et al., "Guiding Global Best Practice in Personalized Nutrition Based on Genetics: The Development of a Nutrigenomics Care Map," *J Acad Nutr Diet* 122, no. 2 (2022): 259–69.
121. D. Roosan et al., "Opportunities to Integrate Nutrigenomics into Clinical Practice and Patient Counseling," *Eur J Clin Nutr* 77, no. 1 (2023): 36–44.
122. World Health Organization, "Physical Activity," (2022), https://www.who.int/news-room/fact-sheets/detail/physical-activity (20/03/2024).
123. H. H. Kyu et al., "Physical Activity and Risk of Breast Cancer, Colon Cancer, Diabetes, Ischemic Heart Disease, and Ischemic Stroke Events: Systematic Review and Dose-Response Meta-Analysis for the Global Burden of Disease Study 2013," *BMJ* 354 (2016): i3857.
124. F. C. Bull et al., "World Health Organization 2020 Guidelines on Physical Activity and Sedentary Behaviour," *Br J Sports Med* 54, no. 24 (2020): 1451–62.
125. E. L. Watts et al., "Association of Leisure Time Physical Activity Types and Risks of All-Cause, Cardiovascular, and Cancer Mortality among Older Adults," *JAMA Netw Open* 5, no. 8 (2022): e2228510.
126. S. A. Newsom et al., "A Single Session of Low-Intensity Exercise Is Sufficient to Enhance Insulin Sensitivity into the Next Day in Obese Adults," *Diabetes Care* 36, no. 9 (2013): 2516–22.
127. R. J. Manders, J. W. Van Dijk, and L. J. van Loon, "Low-Intensity Exercise Reduces the Prevalence of Hyperglycemia in Type 2 Diabetes," *Med Sci Sports Exerc* 42, no. 2 (2010): 219–25.
128. S. F. M. Chastin et al., "How Does Light-Intensity Physical Activity Associate with Adult Cardiometabolic Health and Mortality? Systematic Review with Meta-Analysis of Experimental and Observational Studies," *Br J Sports Med* 53, no. 6 (2019): 370–76.
129. S. Amagasa et al., "Is Objectively Measured Light-Intensity Physical Activity Associated with Health Outcomes after Adjustment for Moderate-to-Vigorous Physical Activity in Adults? A Systematic Review," *Int J Behav Nutr Phys Act* 15, no. 1 (2018): 65.

130. Chastin et al., "How Does Light-Intensity Physical Activity Associate with Adult Cardiometabolic Health and Mortality? Systematic Review with Meta-Analysis of Experimental and Observational Studies."
131. M. S. Tremblay et al., "Sedentary Behavior Research Network (SBRN) – Terminology Consensus Project Process and Outcome," *Int J Behav Nutr Phys Act* 14, no. 1 (2017): 75.
132. P. C. Dempsey et al., "New Global Guidelines on Sedentary Behaviour and Health for Adults: Broadening the Behavioural Targets," *Int J Behav Nutr Phys Act* 17 (2020): 151.
133. K. Wirth et al., "Biomarkers Associated with Sedentary Behaviour in Older Adults: A Systematic Review," *Ageing Res Rev* 35 (2017): 87–111.
134. L. A. Brocklebank et al., "Accelerometer-Measured Sedentary Time and Cardiometabolic Biomarkers: A Systematic Review," *Prev Med* 76 (2015): 92–102.
135. L. B. Sardinha et al., "Sedentary Patterns, Physical Activity, and Cardiorespiratory Fitness in Association to Glycemic Control in Type 2 Diabetes Patients," *Front Physiol* 8 (2017): 262.
136. Bull et al., "World Health Organization 2020 Guidelines on Physical Activity and Sedentary Behaviour."
137. Department of Health and Social Care et al., "UK Chief Medical Officers' Physical Activity Guidelines," (2019), https://assets.publishing.service.gov.uk/government/uploads/system/uploads/attachment_data/file/832868/uk-chief-medical-officers-physical-activity-guidelines.pdf (Accessed 21/05/2024).
138. Australian Government Department of Health, "Australia's Physical Activity and Sedentary Behaviour Guidelines. Fact Sheet: Adults (18–64 Years)," (2021), https://www.health.gov.au/resources/publications/physical-activity-and-sedentary-behaviour-guidelines-adults-18-to-64-years-fact-sheet (Accessed 21/10/2021).
139. U. Ekelund et al., "Does Physical Activity Attenuate, or Even Eliminate, the Detrimental Association of Sitting Time with Mortality? A Harmonised Meta-Analysis of Data from More Than 1 Million Men and Women," *Lancet* 388, no. 10051 (2016): 1302–10.
140. D. P. Bailey and C. D. Locke, "Breaking up Prolonged Sitting with Light-Intensity Walking Improves Postprandial Glycemia, but Breaking up Sitting with Standing Does Not," *J Sci Med Sport* 18, no. 3 (2015): 294–8.
141. M. J. Davies et al., "Management of Hyperglycemia in Type 2 Diabetes, 2018: A Consensus Report by the American Diabetes Association (ADA) and the European Association for the Study of Diabetes (EASD)," *Diabetes Care* 41, no. 12 (2018).
142. Dempsey et al., "New Global Guidelines on Sedentary Behaviour and Health for Adults: Broadening the Behavioural Targets."
143. C. Bouchard and T. Rankinen, "Individual Differences in Response to Regular Physical Activity," *Med Sci Sports Exerc* 33, no. 6 Suppl (2001): S446–51.
144. D. Prud'homme et al., "Sensitivity of Maximal Aerobic Power to Training Is Genotype-Dependent," *Med Sci Sports Exerc* 16, no. 5 (1984): 489–93.
145. Bouchard and Rankinen, "Individual Differences in Response to Regular Physical Activity."
146. C. Pickering and J. Kiely, "Do Non-Responders to Exercise Exist-and If So, What Should We Do About Them?," *Sports Med* 49, no. 1 (2019): 1–7.
147. G. Atkinson and A. M. Batterham, "True and False Interindividual Differences in the Physiological Response to an Intervention," *Exp Physiol* 100, no. 6 (2015): 577–88.
148. T. J. Hrubeniuk et al., "Increasing Aerobic Exercise Intensity Fails to Consistently Improve the Glycemic Response in People Living with Prediabetes or Type 2 Diabetes Mellitus: The Intensity Trial," *Appl Physiol Nutr Metab* 49, no. 6 (2024): 792–804.
149. C. Bouchard et al., "Adverse Metabolic Response to Regular Exercise: Is It a Rare or Common Occurrence?," *PLoS One* 7, no. 5 (2012): e37887.

150. A. J. Hautala et al., "Individual Differences in the Responses to Endurance and Resistance Training," *Eur J Appl Physiol* 96, no. 5 (2006): 535–42.
151. T. A. Churchward-Venne et al., "There Are No Nonresponders to Resistance-Type Exercise Training in Older Men and Women," *J Am Med Dir Assoc* 16, no. 5 (2015): 400–11.
152. S. B. Sisson et al., "Volume of Exercise and Fitness Nonresponse in Sedentary, Postmenopausal Women," *Med Sci Sports Exerc* 41, no. 3 (2009): 539–45.
153. T. N. Mann, R. P. Lamberts, and M. I. Lambert, "High Responders and Low Responders: Factors Associated with Individual Variation in Response to Standardized Training," *Sports Med* 44, no. 8 (2014): 1113–24.
154. A. M. Kiviniemi et al., "Daily Exercise Prescription on the Basis of HR Variability among Men and Women," *Med Sci Sports Exerc* 42, no. 7 (2010): 1355–63.
155. J. S. Skinner et al., "Age, Sex, Race, Initial Fitness, and Response to Training: The Heritage Family Study," *J Appl Physiol (1985)* 90, no. 5 (2001): 1770–6.
156. Mann, Lamberts, and Lambert, "High Responders and Low Responders: Factors Associated with Individual Variation in Response to Standardized Training."
157. M. J. Delmonico et al., "Alpha-Actinin-3 (ACTN3) R577X Polymorphism Influences Knee Extensor Peak Power Response to Strength Training in Older Men and Women," *J Gerontol A Biol Sci Med Sci* 62, no. 2 (2007): 206–12.
158. T. Rankinen et al., "Heritability of Submaximal Exercise Heart Rate Response to Exercise Training Is Accounted for by Nine Snps," *J Appl Physiol (1985)* 112, no. 5 (2012): 892–7.
159. C. Bouchard et al., "Genomic Predictors of the Maximal O_2 Uptake Response to Standardized Exercise Training Programs," *J Appl Physiol (1985)* 110 (2011): 1160–70.
160. N. R. Harvey et al., "Genetic Variants Associated with Exercise Performance in Both Moderately Trained and Highly Trained Individuals," *Mol Genet Genomics* 295, no. 2 (2020): 515–23.
161. M. A. Sarzynski, S. Ghosh, and C. Bouchard, "Genomic and Transcriptomic Predictors of Response Levels to Endurance Exercise Training," *J Physiol* 595, no. 9 (2017): 2931–39.
162. C. E. Marsh et al., "Fitness and Strength Responses to Distinct Exercise Modes in Twins: Studies of Twin Responses to Understand Exercise as a Therapy (Strueth) Study," *J Physiol* 598, no. 18 (2020): 3845–58.
163. J. Tyrrell et al., "Gene-Obesogenic Environment Interactions in the UK Biobank Study," *Int J Epidemiol* 46, no. 2 (2017): 559–75.
164. Y. C. Klimentidis et al., "FTO Association and Interaction with Time Spent Sitting," *Int J Obes (Lond)* 40, no. 3 (2016): 411–6.
165. Y. C. Klimentidis et al., "Association of Physical Activity with Lower Type 2 Diabetes Incidence Is Weaker among Individuals at High Genetic Risk," *Diabetologia* 57, no. 12 (2014): 2530–4.
166. T. O. Kilpelainen et al., "Multi-Ancestry Study of Blood Lipid Levels Identifies Four Loci Interacting with Physical Activity," *Nat Commun* 10, no. 1 (2019): 376.
167. T. Higashibata et al., "Enos Genotype Modifies the Effect of Leisure-Time Physical Activity on Serum Triglyceride Levels in a Japanese Population," *Lipids Health Dis* 11, no. 1 (2012): 150.
168. R. D. Esposti et al., "Influence of Enos Gene Polymorphism on Cardiometabolic Parameters in Response to Physical Training in Postmenopausal Women," *Braz J Med Biol Res* 44, no. 9 (2011): 855–63.
169. N. Grarup et al., "The -250g>a Promoter Variant in Hepatic Lipase Associates with Elevated Fasting Serum High-Density Lipoprotein Cholesterol Modulated by Interaction with Physical Activity in a Study of 16,156 Danish Subjects," *J Clin Endocrinol Metab* 93, no. 6 (2008): 2294–9.

170. C. Bouchard, "Exercise Genomics – A Paradigm Shift Is Needed: A Commentary," *Br J Sports Med* 49, no. 23 (2015): 1492–6.
171. M. A. B. Khan et al., "Epidemiology of Type 2 Diabetes – Global Burden of Disease and Forecasted Trends," *J Epidemiol Glob Health* 10, no. 1 (2020): 107–111.
172. Khan et al., "Epidemiology of Type 2 Diabetes – Global Burden of Disease and Forecasted Trends."
173. R. A. DeFronzo et al., "Type 2 Diabetes Mellitus," *Nat Rev Dis Primers* 1, no. 1 (2015): 15019.
174. F. B. Hu, "Globalization of Diabetes: The Role of Diet, Lifestyle, and Genes," *Diabetes Care* 34, no. 6 (2011): 1249–57.
175. F. P. Cappuccio et al., "Quantity and Quality of Sleep and Incidence of Type 2 Diabetes: A Systematic Review and Meta-Analysis," *Diabetes Care* 33, no. 2 (2010): 414–20.
176. A. Grontved and F. B. Hu, "Television Viewing and Risk of Type 2 Diabetes, Cardiovascular Disease, and All-Cause Mortality: A Meta-Analysis," *JAMA* 305, no. 23 (2011): 2448–55.
177. A. Pan et al., "Rotating Night Shift Work and Risk of Type 2 Diabetes: Two Prospective Cohort Studies in Women," *PLoS Med* 8, no. 12 (2011): e1001141.
178. Khan et al., "Epidemiology of Type 2 Diabetes - Global Burden of Disease and Forecasted Trends."
179. E. Agardh et al., "Type 2 Diabetes Incidence and Socio-Economic Position: A Systematic Review and Meta-Analysis," *Int J Epidemiol* 40, no. 3 (2011): 804–18.
180. Chatterjee, Khunti, and Davies, "Type 2 Diabetes."
181. DeFronzo et al., "Type 2 Diabetes Mellitus."
182. DeFronzo et al., "Type 2 Diabetes Mellitus."
183. Chatterjee, Khunti, and Davies, "Type 2 Diabetes."
184. M. A. Nauck and J. J. Meier, "Incretin Hormones: Their Role in Health and Disease," *Diabetes, Obesity and Metabolism* 20 (2018): 5–21.
185. Nauck and Meier, "Incretin Hormones: Their Role in Health and Disease."
186. S. Li et al., "Adiponectin Levels and Risk of Type 2 Diabetes: A Systematic Review and Meta-Analysis," *JAMA* 302, no. 2 (2009): 179–88.
187. X. Wang et al., "Inflammatory Markers and Risk of Type 2 Diabetes: A Systematic Review and Meta-Analysis," *Diabetes Care* 36, no. 1 (2013): 166–75.
188. A. Lindqvist et al., "Ghrelin Suppresses Insulin Secretion in Human Islets and Type 2 Diabetes Patients Have Diminished Islet Ghrelin Cell Number and Lower Plasma Ghrelin Levels," *Mol Cell Endocrinol* 511 (2020): 110835.
189. Y. Zhou and L. Rui, "Leptin Signaling and Leptin Resistance," *Front Med* 7, no. 2 (2013): 207–22.
190. M. Mars et al., "Fasting Leptin and Appetite Responses Induced by a 4-Day 65%-Energy-Restricted Diet," *Int J Obes (Lond)* 30, no. 1 (2006): 122–8.
191. I. S. Farooqi and S. O'Rahilly, "20 Years of Leptin: Human Disorders of Leptin Action," *J Endocrinol* 223, no. 1 (2014): T63–70.
192. J. M. Friedman, "Leptin and the Endocrine Control of Energy Balance," *Nat Metab* 1, no. 8 (2019): 754–64.
193. T. A. Wadden et al., "Short- and Long-Term Changes in Serum Leptin Dieting Obese Women: Effects of Caloric Restriction and Weight Loss," *J Clin Endocrinol Metab* 83, no. 1 (1998): 214–8.
194. C. Vavruch et al., "Serum Leptin Levels Are Independently Related to the Incidence of Ischemic Heart Disease in a Prospective Study of Patients with Type 2 Diabetes," *Cardiovasc Diabetol* 14 (2015): 62.
195. L. G. F. de Carvalho et al., "Circulating Leptin Levels as a Potential Biomarker in Inflammatory Bowel Diseases: A Systematic Review and Meta-Analysis," *Inflamm Bowel Dis* 27, no. 2 (2021): 169–81.

196. N. Katsiki, D. P. Mikhailidis, and M. Banach, "Leptin, Cardiovascular Diseases and Type 2 Diabetes Mellitus," *Acta Pharmacol Sin* 39, no. 7 (2018): 1176–88.
197. Zhou and Rui, "Leptin Signaling and Leptin Resistance."
198. Mars et al., "Fasting Leptin and Appetite Responses Induced by a 4-Day 65%-Energy-Restricted Diet."
199. M. J. Fowler, "Microvascular and Macrovascular Complications of Diabetes," *Clin Diabetes* 26, no. 2 (2008): 77–82.
200. E. W. Gregg, N. Sattar, and M. K. Ali, "The Changing Face of Diabetes Complications," *Lancet Diabetes Endocrinol* 4, no. 6 (2016): 537–47.
201. N. Sakran et al., "The Many Faces of Diabetes. Is There a Need for Re-Classification? A Narrative Review," *BMC Endocr Disord* 22, no. 1 (2022): 9.
202. Y. Zhang et al., "Combined Lifestyle Factors and Risk of Incident Type 2 Diabetes and Prognosis among Individuals with Type 2 Diabetes: A Systematic Review and Meta-Analysis of Prospective Cohort Studies," *Diabetologia* 63, no. 1 (2020): 21–33.
203. J. Pearson-Stuttard et al., "Trends in Predominant Causes of Death in Individuals with and without Diabetes in England from 2001 to 2018: An Epidemiological Analysis of Linked Primary Care Records," *Lancet Diabetes Endocrinol* 9, no. 3 (2021): 165–73.
204. K. Hemminki et al., "Familial Risks for Type 2 Diabetes in Sweden," *Diabetes Care* 33, no. 2 (2010): 293–7.
205. DeFronzo et al., "Type 2 Diabetes Mellitus."
206. A. I. Young, F. Wauthier, and P. Donnelly, "Multiple Novel Gene-by-Environment Interactions Modify the Effect of FTO Variants on Body Mass Index," *Nat Commun* 7, no. 1 (2016): 12724.
207. Dashti et al., "Gene-Environment Interactions of Circadian-Related Genes for Cardiometabolic Traits."
208. Villegas et al., "Gene-Carbohydrate and Gene-Fiber Interactions and Type 2 Diabetes in Diverse Populations from the National Health and Nutrition Examination Surveys (Nhanes) as Part of the Epidemiologic Architecture for Genes Linked to Environment (Eagle) Study."
209. Zheng et al., "Genetic Variants at PSMD3 Interact with Dietary Fat and Carbohydrate to Modulate Insulin Resistance."
210. Huang et al., "Pcsk7 Genotype Modifies Effect of a Weight-Loss Diet on 2-Year Changes of Insulin Resistance: The Pounds Lost Trial."
211. Zheng et al., "Modulation by Dietary Fat and Carbohydrate of IRS1 Association with Type 2 Diabetes Traits in Two Populations of Different Ancestries."
212. S. R. Colberg et al., "Physical Activity/Exercise and Diabetes: A Position Statement of the American Diabetes Association," *Diabetes Care* 39, no. 11 (2016): 2065–79.
213. R. Patterson et al., "Sedentary Behaviour and Risk of All-Cause, Cardiovascular and Cancer Mortality, and Incident Type 2 Diabetes: A Systematic Review and Dose Response Meta-Analysis," *Eur J Epidemiol* 33, no. 9 (2018): 811–29.
214. J. Del Pozo-Cruz et al., "Replacing Sedentary Time: Meta-Analysis of Objective-Assessment Studies," *Am J Prev Med* 55, no. 3 (2018): 395–402.
215. Brocklebank et al., "Accelerometer-Measured Sedentary Time and Cardiometabolic Biomarkers: A Systematic Review."
216. Evert et al., "Nutrition Therapy for Adults with Diabetes or Prediabetes: A Consensus Report."
217. American Diabetes Association Professional Practice Committee, "2. Classification and Diagnosis of Diabetes: Standards of Medical Care in Diabetes-2022," *Diabetes Care* 45, no. Suppl 1 (2022): S17-S38.
218. E. Ahlqvist et al., "Novel Subgroups of Adult-Onset Diabetes and Their Association with Outcomes: A Data-Driven Cluster Analysis of Six Variables," *Lancet Diabetes Endocrinol* 6, no. 5 (2018): 361–69.

219. R. M. Anjana et al., "Novel Subgroups of Type 2 Diabetes and Their Association with Microvascular Outcomes in an Asian Indian Population: A Data-Driven Cluster Analysis: The Inspired Study," *BMJ Open Diabetes Res Care* 8, no. 1 (2020): 1506.
220. S. B. Cho, S. C. Kim, and M. G. Chung, "Identification of Novel Population Clusters with Different Susceptibilities to Type 2 Diabetes and Their Impact on the Prediction of Diabetes," *Sci Rep* 9, no. 1 (2019): 3329.
221. T. Karpati et al., "Patient Clusters Based on Hbalc Trajectories: A Step toward Individualized Medicine in Type 2 Diabetes," *PLoS One* 13, no. 11 (2018): e0207096.
222. K. Alexandre et al., "Identification of Diabetes Self-Management Profiles in Adults: A Cluster Analysis Using Selected Self-Reported Outcomes," *PLoS One* 16, no. 1 (2021): e0245721.
223. M. C. Amato et al., "Phenotyping of Type 2 Diabetes Mellitus at Onset on the Basis of Fasting Incretin Tone: Results of a Two-Step Cluster Analysis," *J Diabetes Investig* 7, no. 2 (2016): 219–25.
224. Anjana et al., "Novel Subgroups of Type 2 Diabetes and Their Association with Microvascular Outcomes in an Asian Indian Population: A Data-Driven Cluster Analysis: The Inspired Study."
225. Karpati et al., "Patient Clusters Based on Hbalc Trajectories: A Step toward Individualized Medicine in Type 2 Diabetes."
226. Cho, Kim, and Chung, "Identification of Novel Population Clusters with Different Susceptibilities to Type 2 Diabetes and Their Impact on the Prediction of Diabetes."
227. J. C. Arevalo Lorido et al., "Comorbidity in Patients with Type 2 Diabetes Mellitus and Heart Failure with Preserved Ejection Fraction. Cluster Analysis of the Rica Registry. Opportunities for Improvement," *Rev Clin Esp (Barc)* 220, no. 7 (2020): 409–16.
228. Alexandre et al., "Identification of Diabetes Self-Management Profiles in Adults: A Cluster Analysis Using Selected Self-Reported Outcomes."
229. Ahlqvist et al., "Novel Subgroups of Adult-Onset Diabetes and Their Association with Outcomes: A Data-Driven Cluster Analysis of Six Variables."
230. Amato et al., "Phenotyping of Type 2 Diabetes Mellitus at Onset on the Basis of Fasting Incretin Tone: Results of a Two-Step Cluster Analysis."
231. J. M. Dennis et al., "Disease Progression and Treatment Response in Data-Driven Subgroups of Type 2 Diabetes Compared with Models Based on Simple Clinical Features: An Analysis Using Clinical Trial Data," *Lancet Diabetes Endocrinol* 7, no. 6 (2019): 442–51.
232. L. Li et al., "Identification of Type 2 Diabetes Subgroups through Topological Analysis of Patient Similarity," *Sci Transl Med* 7, no. 311 (2015): 311ra174.
233. A. R. Kahkoska et al., "Validation of Distinct Type 2 Diabetes Clusters and Their Association with Diabetes Complications in the Devote, Leader and Sustain-6 Cardiovascular Outcomes Trials," *Diabetes Obes Metab* 22, no. 9 (2020): 1537–47.
234. Karpati et al., "Patient Clusters Based on Hbalc Trajectories: A Step toward Individualized Medicine in Type 2 Diabetes."
235. C. Sudlow et al., "UK Biobank: An Open Access Resource for Identifying the Causes of a Wide Range of Complex Diseases of Middle and Old Age," *PLoS Med* 12, no. 3 (2015): e1001779.
236. J. E. Gerich, "Type 2 Diabetes Mellitus Is Associated with Multiple Cardiometabolic Risk Factors," *Clin Cornerstone* 8, no. 3 (2007): 53–68.
237. K. Faerch et al., "Trajectories of Cardiometabolic Risk Factors before Diagnosis of Three Subtypes of Type 2 Diabetes: A Post-Hoc Analysis of the Longitudinal Whitehall Ii Cohort Study," *Lancet Diabetes Endocrinol* 1, no. 1 (2013): 43–51.

238. A. Hulman et al., "Trajectories of Glycaemia, Insulin Sensitivity and Insulin Secretion in South Asian and White Individuals before Diagnosis of Type 2 Diabetes: A Longitudinal Analysis from the Whitehall Ii Cohort Study," *Diabetologia* 60, no. 7 (2017): 1252–60.
239. D. Noble et al., "Risk Models and Scores for Type 2 Diabetes: Systematic Review," *BMJ* 343 (2011): d7163.
240. N. Nanayakkara et al., "Impact of Age at Type 2 Diabetes Mellitus Diagnosis on Mortality and Vascular Complications: Systematic Review and Meta-Analyses," *Diabetologia* 64, no. 2 (2021): 275–87.
241. S. Zoungas et al., "Impact of Age, Age at Diagnosis and Duration of Diabetes on the Risk of Macrovascular and Microvascular Complications and Death in Type 2 Diabetes," *Diabetologia* 57, no. 12 (2014).
242. S. M. Shamshirgaran et al., "Age Differences in Diabetes-Related Complications and Glycemic Control," *BMC Endocr Disord* 17, no. 1 (2017): 25.
243. M. Lugner et al., "Identifying Top Ten Predictors of Type 2 Diabetes through Machine Learning Analysis of UK Biobank Data," *Sci Rep* 14, no. 1 (2024): 2102.
244. K. Li et al., "Causal Associations of Waist Circumference and Waist-to-Hip Ratio with Type II Diabetes Mellitus: New Evidence from Mendelian Randomization," *Mol Genet Genomics* 296, no. 3 (2021): 605–13.
245. E. Ahlqvist et al., "Novel Subgroups of Adult-Onset Diabetes and Their Association with Outcomes: A Data-Driven Cluster Analysis of Six Variables," *Lancet Diabetes Endocrinol* 6, no. 5 (2018): 361–69.
246. N. Gray et al., "The Relationship between BMI and Onset of Diabetes Mellitus and Its Complications," *South Med J* 108, no. 1 (2015): 29.
247. W. S. Leslie et al., "The Diabetes Remission Clinical Trial (Direct): Protocol for a Cluster Randomised Trial," *BMC Fam Pract* 17, no. 1 (2016): 20.
248. G. Thom et al., "Predictors of Type 2 Diabetes Remission in the Diabetes Remission Clinical Trial (Direct)," *Diabet Med* 38, no. 8 (2021): e14395.
249. A. Jan and C. B. Weir, "BMI Classification Percentile and Cut-Off Points," in *StatPearls* (Treasure Island, FL: StatPearls Publishing, 2021).
250. G. Whitlock et al., "Prospective Studies Collaboration: Body-Mass Index and Cause-Specific Mortality in 900 000 Adults: Collaborative Analyses of 57 Prospective Studies," *Lancet* 373, no. 9669 (2009): 1083–96.
251. H. R. Banack et al., "Is BMI a Valid Measure of Obesity in Postmenopausal Women?," *Menopause* 25, no. 3 (2018): 307–13.
252. S. Humphreys, "The Unethical Use of BMI in Contemporary General Practice," *Br J Gen Pract* 60, no. 578 (2010): 696–7.
253. I. Janssen, P. T. Katzmarzyk, and R. Ross, "Waist Circumference and Not Body Mass Index Explains Obesity-Related Health Risk," *Am J Clin Nutr* 79, no. 3 (2004): 379–84.
254. E. Krishnan et al., "Hyperuricemia in Young Adults and Risk of Insulin Resistance, Prediabetes, and Diabetes: A 15-Year Follow-up Study," *Am J Epidemiol* 176, no. 2 (2012): 108–16; Janssen, Katzmarzyk, and Ross, "Waist Circumference and Not Body Mass Index Explains Obesity-Related Health Risk."
255. DeFronzo et al., "Type 2 Diabetes Mellitus."
256. N. Sasaki et al., "Association of Insulin Resistance, Plasma Glucose Level, and Serum Insulin Level with Hypertension in a Population with Different Stages of Impaired Glucose Metabolism," *J Am Heart Assoc* 9, no. 7 (2020): e015546.
257. T. S. Hannon et al., "Review of Methods for Measuring Beta-Cell Function: Design Considerations from the Restoring Insulin Secretion (Rise) Consortium," *Diabetes Obes Metab* 20, no. 1 (2018): 14–24.

258. Sasaki et al., "Association of Insulin Resistance, Plasma Glucose Level, and Serum Insulin Level with Hypertension in a Population with Different Stages of Impaired Glucose Metabolism."
259. M. Lind et al., "The Association between Hba1c, Fasting Glucose, 1-Hour Glucose and 2-Hour Glucose During an Oral Glucose Tolerance Test and Cardiovascular Disease in Individuals with Elevated Risk for Diabetes," *PLoS One* 9, no. 10 (2014): e109506.
260. World Health Organization, "Use of Glycated Haemoglobin (Hba1c) in Diagnosis of Diabetes Mellitus: Abbreviated Report of a WHO Consultation," World Health Organization, (2011), https://www.ncbi.nlm.nih.gov/books/NBK304267/ (Accessed 24/03/2024).
261. G. Kaiafa et al., "Is Hba1c an Ideal Biomarker of Well-Controlled Diabetes?," *Postgrad Med J* 97, no. 1148 (2021): 380–83.
262. Lugner et al., "Identifying Top Ten Predictors of Type 2 Diabetes through Machine Learning Analysis of UK Biobank Data."
263. K. R. Klein and J. B. Buse, "The Trials and Tribulations of Determining Hba(1c) Targets for Diabetes Mellitus," *Nat Rev Endocrinol* 16, no. 12 (2020): 717–30.
264. V. Boudreau et al., "Screening for Cystic Fibrosis-Related Diabetes: Matching Pathophysiology and Addressing Current Challenges," *Can J Diabetes* 40, no. 5 (2016): 466–70.
265. J. R. Petrie, T. J. Guzik, and R. M. Touyz, "Diabetes, Hypertension, and Cardiovascular Disease: Clinical Insights and Vascular Mechanisms," *Can J Cardiol* 34, no. 5 (2018): 575–84.
266. C. A. Emdin et al., "Blood Pressure Lowering in Type 2 Diabetes: A Systematic Review and Meta-Analysis," *JAMA* 313, no. 6 (2015): 603–15.
267. M. Nazarzadeh et al., "Blood Pressure Lowering and Risk of New-Onset Type 2 Diabetes: An Individual Participant Data Meta-Analysis," *Lancet* 398, no. 10313 (2021): 1803–10.
268. L. Gao et al., "Association of Apolipoproteins A1 and B with Type 2 Diabetes and Fasting Blood Glucose: A Cross-Sectional Study," *BMC Endocr Disord* 21, no. 1 (2021): 59.
269. T. G. Richardson et al., "Effects of Apolipoprotein B on Lifespan and Risks of Major Diseases Including Type 2 Diabetes: A Mendelian Randomisation Analysis Using Outcomes in First-Degree Relatives," *Lancet Healthy Longev* 2, no. 6 (2021): e317-e26.
270. G. Reyes-Soffer et al., "Lipoprotein(a): A Genetically Determined, Causal, and Prevalent Risk Factor for Atherosclerotic Cardiovascular Disease: A Scientific Statement from the American Heart Association," *Arterioscler Thromb Vasc Biol* 42, no. 1 (2022): e48-e60.
271. Y. Zhang et al., "Lipoprotein (a) Predicts Recurrent Worse Outcomes in Type 2 Diabetes Mellitus Patients with Prior Cardiovascular Events: A Prospective, Observational Cohort Study," *Cardiovasc Diabetol* 19, no. 1 (2020): 111.
272. C. Lamina and N. C. Ward, "Lipoprotein (a) and Diabetes Mellitus," *Atherosclerosis* 349 (2022): 63–71.
273. J. Zhao et al., "Triglyceride Is an Independent Predictor of Type 2 Diabetes among Middle-Aged and Older Adults: A Prospective Study with 8-Year Follow-Ups in Two Cohorts," *J Transl Med* 17, no. 1 (2019): 403.
274. A. S. Alexopoulos et al., "Triglycerides: Emerging Targets in Diabetes Care? Review of Moderate Hypertriglyceridemia in Diabetes," *Curr Diab Rep* 19, no. 4 (2019): 13.
275. G. F. Watts, E. M. Ooi, and D. C. Chan, "Demystifying the Management of Hypertriglyceridaemia," *Nat Rev Cardiol* 10, no. 11 (2013): 648–61.
276. S. C. Thambiah and L. C. Lai, "Diabetic Dyslipidaemia," *Pract Lab Med* 26 (2021): e00248.

277. Thambiah and Lai, "Diabetic Dyslipidaemia."
278. S. Kanmani et al., "Association of C-Reactive Protein with Risk of Developing Type 2 Diabetes Mellitus, and Role of Obesity and Hypertension: A Large Population-Based Korean Cohort Study," *Sci Rep* 9, no. 1 (2019): 4573.
279. P. J. Rousseeuw, "Silhouettes: A Graphical Aid to the Interpretation and Validation of Cluster Analysis," *J Comput Appl Math* 20 (1987).
280. C. Hennig, "Cluster-Wise Assessment of Cluster Stability," *Comput Stat Data Anal* 52, no. 1 (2007): 53–65.
281. Hennig, "Cluster-Wise Assessment of Cluster Stability."
282. F. Pedregosa et al., "Scikit-Learn: Machine Learning in Python," *J Mach Learn Res* 12, no. 85 (2011): 2825–30.
283. Ahlqvist et al., "Novel Subgroups of Adult-Onset Diabetes and Their Association with Outcomes: A Data-Driven Cluster Analysis of Six Variables."

2 Understanding the relation between movement and health using statistics and machine learning

INTRODUCTION

Physical behaviours, including physical activity (PA) and sedentary time, are crucial determinants of health, significantly influencing the risk of developing lifestyle-related conditions such as type 2 diabetes (T2D), cardiovascular disease (CVD), and obesity. Despite extensive research, many aspects of the relationship between physical behaviours and health remain unclear. This chapter starts by providing a comprehensive review of the known effects of physical behaviours on health, with a particular focus on T2D. The well-documented benefits of PA for glucose control and overall health are discussed, as well as the less understood but significant effects of sedentary behaviour (SB). The interaction between different intensities of physical activities and sedentary time is explored, highlighting gaps in the current evidence base and the need for further research.

Next, the chapter examines various methods for collecting data on physical behaviours. Traditional self-report measures, such as activity logs and questionnaires, are discussed alongside objective measurement tools like accelerometers and pedometers. The advantages and limitations of these methods are evaluated, with an emphasis on the importance of accurate and reliable data collection for epidemiological research. The chapter then delves into the statistical techniques commonly used to analyse the relationship between physical behaviours and health outcomes. It covers linear regression, isometric log-ratio (ILR) transformation, and isotemporal substitution models, explaining how each method addresses the unique challenges posed by physical behaviour data. The strengths and limitations of these techniques are discussed in the context of their ability to uncover meaningful associations and inform public health guidelines.

Two case studies are presented to illustrate the practical application of these concepts. The first case study uses Hidden Semi-Markov Models (HSMMs) to classify human activities from accelerometer data. This approach demonstrates the potential

DOI: 10.1201/9781003464983-2

of machine learning (ML) algorithms to improve the accuracy of activity classification and the insights derived from such data.

The second case study integrates principles of personalised medicine and epidemiology to investigate subgroup-specific associations between types of T2D and physical behaviours. By building on the results presented in the Chapter 1 case study, which identified distinct clusters of T2D based on cardiometabolic markers, this study explores how different physical behaviour patterns relate to health outcomes within each cluster. The findings underscore the importance of tailored interventions in managing T2D and highlight the potential for personalised lifestyle medicine to improve patient outcomes.

In conclusion, this chapter aims to enhance the understanding of the complex interplay between physical behaviours and health. By combining advanced data collection methods, robust statistical techniques, and personalised medicine approaches, it provides a comprehensive framework for studying and improving health outcomes related to PA and SB.

PHYSICAL BEHAVIOURS AND HEALTH

As discussed in Chapter 1, time spent on PA and SBs is a risk factor that greatly affects health and the risk of developing metabolism-related conditions, including obesity, T2D, and CVD. There is increasing evidence around the negative effects of inactivity; therefore, the first-line therapy to improve glucose control in T2D is lifestyle management, which includes PA promotion[1] as well as reduction of SB. While the benefits of PA are well known, less evidence is available regarding the effects of SB and its interaction with other activities, especially low-intensity activities. SB is defined as any activity below 1.5 Metabolic Equivalents of Task (MET) in either a lying, reclining, or sitting position,[2] and is unfavourably associated with several T2D biomarkers independent of moderate-to-vigorous physical activity (MVPA).[3,4] Despite some countries introducing guidelines on SB, such as in the UK[5] and Australia,[6] a recent study[7] summarising the research behind the newest World Health Organization (WHO) guidelines[8] highlighted the lack of sufficient evidence to set time-based recommendations regarding SB. Additionally, the study emphasised the need to establish alternative ways to offset the damaging effects of SB, given that the high levels of MVPA (>300 minutes/week) needed to reduce the mortality risk—as suggested by the current evidence—may be unfeasible for a large part of the population. Notably, in 2020, Canada introduced 24-hour movement guidelines for adults[9] suggesting spending no more than 8 hours in SBs, including no more than 3 hours of recreational screen time. The guidelines also suggest spending several hours doing light physical activities, including standing, and at least 150 minutes per week of MVPA. Australia,[10] New Zealand,[11] South Africa,[12] and the WHO[13] have released 24-hour movement guidelines, which include time-based recommendations on SB for the early years and children, but have not done so for adults.

In order to understand which lifestyle changes to implement, evidence needs to be provided on how to optimise activity type and intensity in a manner that is both beneficial and feasible. Despite PA being a well-known factor impacting several health outcomes, most of the available evidence focuses on MVPA and its benefits on health—much less clear is the effect of low-intensity physical activity (LPA) on cardiometabolic conditions and their associated biomarkers, and whether LPA can counterbalance the negative effects of SB. LPA is defined as any activity performed under 3 METs, and includes casual walking, household chores, and the activities of daily living.[14] Although several works document the benefits of LPA for markers of T2D,[15,16,17] many studies analysing the cross-sectional association between SB and diabetes markers have not included LPA in the analyses; the high collinearity between the two variables prevents them from being included simultaneously in regression models. For example, a review by Chastin et al.[18] found that LPA improves cardiometabolic health, but did not include information on the relationship with SB. Moreover, Amagasa et al.[19] reviewed the effects of LPA on cardiometabolic biomarkers and found positive associations independently of MVPA, but could not find evidence of the combined effects of time spent in PA and SB. Studies using the isotemporal substitution model (ISM)[20] or a compositional transformation[21] have been able to include LPA in the models and reported benefits associated with the reallocation of SB to LPA. Although alternative statistical methods to simple linear regression have allowed inclusion of different types of physical behaviours, the evidence is still not sufficient to provide specific recommendations around how much time one should spend in different activity types. Several reviews[22,23,24,25] summarised the evidence around the reallocation of time between SB and PA in the adult population, but did not address whether there is a minimum and maximum allocation time of LPA and MVPA for which such benefits could be observed.

Different time allocations and activity compositions impact the magnitude of the reduction of T2D markers. When considering several studies around the effect of SB on T2D biomarkers, there is some evidence supporting the case for time-based guidelines to reduce prolonged periods of sedentary time in the 18–64 year-old healthy population. Gibbs et al.[26] reported a 4.8% increase in fasting insulin for every additional hour of SB, and Celis-Morales et al.[27] found that the highest SB quartile had a 6.2% increase in insulin compared to the lowest quartile of SB. According to Buman et al.,[28] substituting 30 minutes of SB with LPA reduced insulin by 2.4%, and substituting with MVPA reduced insulin by 14.5%. These effect sizes are comparable to the effect of an LPA intervention,[29] which produced an 18.2% reduction after three months in high-risk patients for T2D. Additionally, a 20% difference in the area under the insulin response curve was found to be associated with a 10% difference in coronary mortality risk,[30] thus suggesting that SB reallocations could have a clinically meaningful effect. However, other studies reported much smaller percentage reductions in insulin: Carson et al.[31] reported a 0.03% change for every additional hour of SB; Healy et al.[32] and Lynch et al.[33] found 0.32% and 0.34% insulin differences between the top and bottom quartiles of SB (with a difference in SB of 2.3 and 2.1 hours/day), respectively. Notably, a recent analysis of the Maastricht Study[34] considering the totality of behaviours that compose the 24-hour day, calculated the time-use composition most optimally associated with metabolic risk factors. The authors found that spending 6 hours sitting, 5 hours and 10 minutes standing, 2 hours and 10 minutes in LPA, 2 hours and 10 minutes in MVPA, and 8 hours and 20 minutes sleeping resulted in a beneficial mix. These results show that not only MVPA and LPA, but also standing, is a beneficial activity to counteract the negative effects of sedentary time.

Despite the evidence documenting the deleterious effects of high SB, a review on the interaction between subjectively measured MVPA and SB[35] found that adults who engaged in MVPA for 60–75 minutes per day did not have an increase in mortality risk even with sitting times greater than 8 hours/day. However, 60–75 minutes per day of MVPA may not be achievable by the general population, and displacing SB with LPA, or a mixture of MVPA and LPA, could be a more feasible option. Supporting the case for displacing SB with longer bouts of LPA rather than shorter MVPA bouts, a randomised crossover trial[36] found that substituting one hour of daily sitting with MVPA (cycling) does not improve insulin or glucose, while replacing 6 hours/day of sitting with 4 hours of leisurely walking and 2 hours of standing has significant beneficial effects on both insulin and glucose.

However, despite there being some evidence of a cut-off between 7–9 hours/day of SB, with one recent meta-analysis finding a threshold of 9.5 hours for all-cause mortality,[37] more data points are required to support this value, specifically for T2D biomarkers. The current evidence is still deemed insufficient to support more specific guidelines around how to replace time spent sedentary. Although there is an extensive body of research on the topic, a significant challenge in compiling robust evidence lies in the heterogeneity of methods employed to analyse the associations between physical behaviours and health. The findings obtained with different analytical approaches can be in disagreement,[38] and even systematic reviews can reach different conclusions. For example, a systematic review[39] found no association between SB and fasting glucose, while another[40] did, likely due to their inclusion of unadjusted data in their meta-analysis. Standardising statistical methods is essential to enhance the translation of research into public health guidelines and policy. This standardisation will benefit not only future research but also future policymakers by emphasising the importance of selecting appropriate methods for the collection and analysis of data on physical behaviours.

In summary, current guidelines focus on moderate and vigorous PA, but lower-intensity activities can also benefit T2D markers. More prolonged but less intense exercise may be easier for the elderly or impaired, while short but more intense activities could be as beneficial for those who cannot engage in prolonged sessions. Future research should analyse the effects of different bout lengths on a wide array of metabolic biomarkers to provide the data necessary to develop new and quantitative guidelines on SB and LPA, alongside the already present MVPA recommendations.

ANALYSING HUMAN MOVEMENT DATA

Analysing human movement data is a critical aspect of understanding the relationship between PA and health outcomes. This process involves collecting accurate and comprehensive data on physical behaviours and employing appropriate analytical methods to interpret this information. Data collection methods range from traditional self-report questionnaires to objective measurement devices like accelerometers, pedometers, and smartphones, which offer precise and continuous tracking of human activity.

Once collected, the data must be accurately classified and analysed using robust statistical methods and algorithms to uncover meaningful patterns and insights. Machine learning algorithms, such as decision trees, random forests, and deep learning models, are increasingly being used to classify accelerometer data, providing insights into movement patterns. However, the heterogeneity of data collection methods, statistical methods, and classification algorithms employed in research studies poses significant challenges to compiling a coherent body of evidence. Standardising these methods is crucial to improving the translation of research into public health guidelines and policy.

This section discusses the process of analysing human movement data, which requires careful consideration of data collection methods and classification algorithms. By understanding the nuances of data collection and analysis, researchers can better elucidate the complex relationships between physical behaviours and health, ultimately contributing to more effective public health guidelines and policies.

Data collection

Various methods exist to collect data on physical behaviours, each offering advantages and limitations. Arguably the most common way to measure physical behaviours is through self-reported measures, such as activity logs and recall questionnaires, which are considered practical, low-cost, and easy to use.[41] Various questionnaires exist to collect data on physical behaviours, with the International Physical Activity Questionnaire[42] (IPAQ) being the most notable and widely used. It contains 31 questions, or 9 in the short-form version, on how much time people spend sitting, in light, moderate, and vigorous activity in a typical week. Scores on the IPAQ can then be used to estimate total energy expenditure via conversion of activities to METs. Despite their convenience, self-reported methods are susceptible to high variability and recall bias,[43] especially in certain populations like young individuals or elderly adults. Interestingly, PA data collected through questionnaires and accelerometer is more closely aligned in fitter individuals, while less fit individuals tend to over-report their PA levels.[44,45]

As opposed to subjective measures, objective measures—accelerometers, pedometers, heart rate monitors, and other sensors—show less variability and allow a more reliable assessment of physical behaviours. Although pedometers, which measure total steps, are a popular tool, accelerometers allow us to examine activity type, duration, and intensity, and thus are preferred when examining the associations between health outcomes and activity types. Accelerometers are usually worn on the hip, wrist, or chest and detect incidences of acceleration that are interpreted as bodily movements. Incidences of acceleration are recorded as "counts" that are logged over a pre-specified time epoch and can be used to determine the intensity of activity. Wearable accelerometers have been used as an effective tool in various research settings, from measuring the impact of physical behaviours on health,[46] to pre-operative evaluations.[47] However, despite the increased accuracy obtainable with wearable devices, they present a disadvantage: they may affect a person's behaviour, who can alter their activity levels due to the novelty of the device.[48] Additionally, while accelerometers provide accurate and reliable measurements of physical behaviours, the lack of standard guidelines around accelerometer type, wear location, epoch length, cut-points, and other parameters introduces a layer of subjectivity in the data collection that is not present for self-reported methods. Methodological decisions around device-based physical behaviour measurement are also associated with differences in participant recruitment, adherence to wear instructions, and data loss.[49] For example, waist-worn accelerometers are associated with lower compliance compared to wrist-worn devices.[50] Finally, the choice of device can impact activity estimation, potentially reducing the likelihood of fair comparisons between studies. An assessment of the disparity between various commercially available software for processing PA data from accelerometers showed large differences between software and thresholds to differentiate between activities.[51] This issue is particularly problematic as it hampers systematic reviews of the literature, thereby slowing the development of guidelines around physical behaviours for health improvement. As a consequence, researchers have called for the urgent development of gold standards to validate different devices and their associated software.

Studies have indicated that self-reported data on PA and SB may not always align with objectively measured data,[52,53] underscoring the importance of utilising both subjective and objective measures to accurately capture physical behaviours.

The discrepancy between cut-offs found with device-measured and self-reported measures of SB is likely due to participants underestimating the time spent in SB: a recent meta-analysis quantified the difference between device-measured and self-reported measures of SB as ~1.74 hours/day.[54] Inconsistencies in data collection methods make it more challenging to systematically summarise the available evidence to inform public health guidelines. For example, a meta-analysis[55] found evidence of an accelerometer-measured 9-hour cut-off of SB—adjusted for MVPA—after which hazard ratios of all-cause mortality increased. However, for studies measuring SB with subjective methods, the cut-off was around 7 hours; a similar cut-off (6–8 hours) for subjective SB was also found by Patterson et al.[56]

Selecting the most appropriate measurement tool is crucial to obtaining robust results. Factors to consider when choosing a measurement technique include feasibility and practicality, as well as validity, reliability, and sensitivity in the population being studied.[57]

Methods for accelerometer data classification

Unlike methods such as questionnaires and pedometers that provide data readily usable in associative analyses, accelerometers require the processing and classification of raw data into meaningful categories of physical behaviours.

The most common method to classify human activity is through the cut-off or thresholding approach. The method requires the development of an in-lab protocol to measure the acceleration magnitude registered by a specific device while engaging in activities of varied intensities. The acceleration magnitude is then regressed against MET values for the recorded activities to obtain device-specific cut-off values. Conventionally, two cut-offs of 1.5 and 3 METs are used to split activity into sedentary, light intensity, and moderate-to-vigorous intensity.[58,59,60,61,62] Counts per minute (cpm) may also be used to classify different types of activity by applying thresholds to associate ranges of cpm to different activities, with more intense exercise being associated with higher cpm. Counts are a unit of measurement representing the amount of activity detected by an accelerometer and are usually generated by the device manufacturer through a proprietary algorithm.

Despite their widespread use, the thresholding method presents several drawbacks. As cut-off points are developed using PA protocols carried out in laboratory settings, they are not reflective of free-living conditions. Calibration studies are required practically for every study, as applying established cut-offs could create biased results since they were validated on different populations, measuring devices, and activity types. Watson et al.[63] found that, even when using the same accelerometer device, the cut-off points range from 191 to 2,743 cpm for moderate-intensity activity and from 4,945 to 7,526 cpm for vigorous-intensity activity, depending on whether the calibration was done on laboratory or free-living activities, reflecting the poor generalisability of the cut-offs approach.

Machine learning methods offer several advantages over cut-off methods for accelerometer data classification. Machine learning algorithms can capture more complex dependencies and nonlinearities in accelerometer data to improve activity segmentation.[64] Supervised machine learning has been successfully applied to accelerometer data to recognise activities. Willetts et al.[65] used Random Forests paired with a Hidden Markov Model (HMM) to classify activities in the UK Biobank and achieved high classification accuracy (≥90%). Farrahi et al.[66] used a Recurrent Neural Network to classify accelerometer data as sleep, SB, LPA, and MVPA, which achieved significantly higher performance (16–30% increase) compared to other machine learning methods, including Random Forests, Support-Vector Machines, and Naïve Bayes. Other deep learning methods, namely deep feed-forward neural networks and deep recurrent neural networks, have been successfully employed by Nunavath et al.[67] to accurately classify specific activities such as descending stairs or sitting on a chair. More recently, Farrahi et al.[68] showed that using a deep convolutional autoencoder paired with K-means clustering to classify activities from accelerometer data represented as images allows segmenting activities on intensity as well as timing.

However, supervised learning requires a large amount of labelled data to train the models, which can be hard to obtain in free-living settings. It could require the participants to complete a very detailed activity log, which nonetheless does not guarantee the accuracy of the labelled set due to recall bias. Willetts et al.[69] used a ground-truth set of labelled data acquired with a camera to train their model, which is extremely resource-intensive for this study and would require participants to wear a camera for several days, in addition to the accelerometer. Alternative solutions include training the model on laboratory-acquired acceleration data, but the prediction accuracy of the classification model can be reduced by 20–30% when cross-validated on free-living acceleration data.[70] Furthermore, using existing datasets to train a model is not feasible, given the heterogeneity in population, device type, and device placement. A further problem related to supervised methods is that they do not capture the time dependence of the data,[71] although temporal dependency can be incorporated using HMMs.[72]

Self-supervised learning can mitigate the reliance on vast labelled datasets, and transfer learning allows the use of open-source datasets to pre-train a model, which is later fine tuned on the population of interest. These concepts have been employed by Creagh et al.[73] who used a deep convolutional neural network to estimate daily activity patterns in people with rheumatoid arthritis and reported improved performance over random forests. For a systematic inventory of self-supervised activity recognition using wearables, the reader is referred to Haresamudram et al.[74]

Unsupervised learning methods offer a viable alternative to the supervised and semi-supervised methods outlined above and have the advantage of not requiring labelled data. Clustering methods are at the heart of unsupervised learning, and common techniques such as K-means,[75,76] K-medoids,[77] and Mixture Models[78] have been applied to human activity recognition. However, clustering techniques present certain limitations in the context of accelerometer data, primarily due to the assumption of data independence, which results in the loss of temporal dependency inherent to accelerometer data and human activity. When analysing accelerometer data, preserving temporal dependency is crucial for distinguishing between activities that might otherwise appear similar. For instance, sensor data may indicate that fidgeting at a desk closely resembles washing dishes in terms of movement patterns. However, the former is typically preceded by a period of sitting, while the latter is more likely to follow a brief episode of walking (e.g., moving towards the sink). Retaining time dependency thus allows for more accurate differentiation between activities based on their preceding and subsequent behaviours.

HMMs solve the time dependency issue by representing and learning the data through the exploitation of their sequential characteristics.[79] They have been found to outperform both K-means and Gaussian Mixture Models when used for the classification of activities recorded in laboratory settings.[80] However, when recording data in free-living conditions, the complexity of activities encountered can impede the performance of HMMs. This limitation arises because HMMs implicitly model state durations using a geometric distribution, which is unlikely to be appropriate for accurately modelling activity durations.[81] State duration is another crucial feature for human activity classification, particularly when activities share similar acceleration profiles but vary in duration, significantly affecting the model's prediction accuracy.[82]

HSMMs address the aforementioned issues by modelling the classification problem as an HMM, while incorporating an explicit distribution for the state duration, which corresponds to the number of observations emitted by each state. This represents the primary distinction between HSMMs and HMMs, as the latter emit only a single observation per state and implicitly model state duration through self-transitions.

HSMM have been used to segment accelerometer data by Van Kuppevelt et al.[83] who, however, found only a small correlation between HSMM-inferred states and states found with the cut-off method. This may be due to the study not performing model selection by testing different feature spaces and other parameters, such as duration and observation priors hyperparameters. More recently, Thornton et al.[84] showed that activity segmentation through HSMM has better clinical utility in children than the thresholding approach, as the estimates of PA produced by the HSMM method correlated more strongly with children's developmental capacity compared to the cut points estimates. Nonetheless, the potential of HSMM inference remains largely unexplored, and the feasibility of HSMM in epidemiological research remains unclear.

STATISTICAL METHODS FOR PHYSICAL BEHAVIOURS EPIDEMIOLOGY

Uncovering associations between physical behaviours and health outcomes requires statistical methodologies capable of handling the characteristics of human activity data. This section explores three key statistical approaches utilised in this context: linear regression, ILR transformation, and isotemporal substitution models. Linear regression provides a foundational framework for investigating direct relationships between physical behaviours and health markers; however, it presents important drawbacks in the context of activity data. Compositional data analysis (CoDA) represents a more appropriate methodology for the analysis of activity data, where the relative distribution of activities matters more than absolute values. Isotemporal substitution analysis offers insights into the health impact of reallocating time between different physical behaviours, while combining isotemporal substitution with CoDA allows the assessment of the impact of substituting activities without making unfitting assumptions around movement data. Choosing the most appropriate method is crucial for obtaining valid, repeatable, and actionable insights.

Linear regression modelling

Linear regression models are used to investigate the effects of explanatory variables $x_1,\&,x_k$ on a continuous response variable y. A general linear model is given by the following equation:

$$y = \beta_0 + \beta_1 x_1 + \ldots + \beta_k x_k + e \tag{2.1}$$

where β_0 is the intercept and $\beta_1,\ldots,\beta_k$ are the regression coefficients for $x_1,\ldots,x_k$, estimated via ordinary least squares.[85] The regression coefficients describe the strength of the effect of the explanatory variables on the response variable. For example, if we wanted to assess the effect of SB (x_1) on body mass index (BMI) (y), we would fit a linear regression model (assuming the relationship between SB and BMI was, in fact, linear) where β_1 represents the effect of SB on BMI, indicating the increase in BMI for each unit increase in SB. This example is illustrated in Figure 2.1. The coefficients are estimated on the sample data, and estimations of

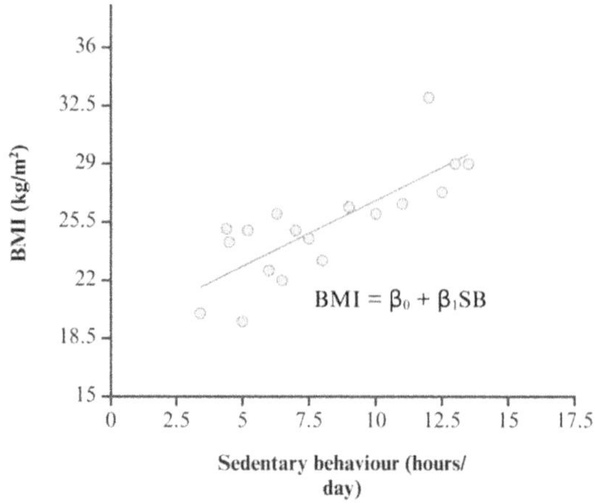

FIGURE 2.1 Illustrative example of linear regression between sedentary behaviour (SB) and body mass index (BMI).

uncertainties, such as the 95% confidence interval and p-value, are conventionally provided alongside the estimated coefficient. If the measures of uncertainty indicate statistical significance, the independent variable x_i has an effect on the dependent variable y.

Linear regression is arguably the most common method to assess the effects of physical behaviours on health outcomes, but it presents several issues when used to analyse activity data. The first practical issue associated with this method is the choice of time increments to use in the analysis. While in the literature 30 minutes seems to be the most common bout length, other lengths ranging from 10 to 120 minutes have been used, thus changing the magnitude of the effect and complicating the review of available evidence due to method heterogeneity. Another issue is the collinearity of time-based physical behaviours. As a result, daily time spent in different activities cannot be simultaneously included in the models, thus preventing the adjustment of the examined behaviour for all the others. Finally, the linearity assumptions impose a tough constraint on the validity of the analysis. This requires special consideration in the case of physical behaviours, which are not linear and bounded in the (0, 24) hours range. Consequently, a change in one of these movement-related behaviours will result in a net equal and opposite change in other behaviours. For these reasons, in 2018 the GRANADA consensus[86] suggested abandoning the use of linear regression when analysing physical behaviours.

Compositional analysis with isometric log-ratio transformation

The field of behavioural epidemiology is increasingly acknowledging the limitations of considering physical behaviours as independent entities, and a new paradigm—the time-use epidemiology paradigm[87]—is paving its way in this field of research.

Given that the time spent in physical behaviours (including sleep) is constrained to a total of 24 hours per day, the proportions of time spent sleeping, in SB, and in different types of PA are perfectly collinear. Additionally, the constant sum means that every change in time spent in one behaviour results in a corresponding opposite change of the time spent in the other behaviours. This type of data can be modelled through CoDA, which requires special mathematical considerations. Due to the high collinearity inherent to compositional data, applying linear regression would incur the aforementioned issues. As a result, the ILR transformation can be used to express a given behaviour relative to the time spent in the rest of the behaviours, thereby eliminating the collinearity issue and allowing the use of linear regression.

In a compositional framework, time spent in different activities is considered a relative proportion of the overall time budget (24 hours), such that the vector $x = [x_1, x_2, \ldots, x_D] \in R^D$, with D being the number of activities and with $\sum_{i=1}^{D} x_i = C$, is constrained by the closure constant $C = 24$ hours. The closure constant implies multicollinearity among the activities, and thus conventional statistical methods cannot be employed with compositional data. The ILR transformation maps the data from the constrained simplex space to the unconstrained real space, which allows for the application of regression. Therefore, the transformation $z = ILR(x)$ is applied to the accelerometer data as follows:

$$z_i = \sqrt{\frac{D-\mathrm{i}}{D-i+1}} ln \frac{x_i}{\sqrt[D-i]{\prod_{j=i+1}^{d} x_j}} \quad \text{with } \mathrm{i} = 1, \ldots, D-1 \tag{2.2}$$

where D is the number of behaviours considered, and b is the amount of time spent in a certain behaviour. For example, to transform SB into a three-composition split of the day in order to assess its effect on the target outcome, we would have the following:

$$D = 3, x_1 = SB,\ x_2 = LIPA,\ x_3 = MVPA \tag{2.3}$$

$$z_1 = \sqrt{\frac{2}{3}} ln \frac{SB}{\sqrt[2]{LIPA \times MVPA}} \tag{2.4}$$

$$z_2 = \sqrt{\frac{1}{2}} ln \frac{LIPA}{\sqrt[1]{MVPA}} \tag{2.5}$$

In the analyses, only regression coefficients for z_1 are reported, since the regression coefficient β_1 for the first ILR coordinate z_1 represents the strength of the association between the chosen activity and the outcome, while z_{1+i} cannot be interpreted in a meaningful way. For a more detailed explanation of the ILR, the reader is referred to Dumuid et al.[88] and Chastin et al.[89]

Isotemporal substitution and partition models

The ISM was developed by Mekary et al.[90] on the basis that total hours in a day are finite, and participating in one activity results in not participating in another. Simply, the ISM is a multiple regression model in which all behaviours of interest except one are used as explanatory variables in order to estimate the effect of replacing one activity with another for the same amount of time. An example ISM with three activities is expressed as:

$$y = \beta_1 activity_1 + \beta_2 activity_2 + \beta_3 activity_3 + \beta_4 total\ activity \quad (2.6)$$

where $\beta_1 - \beta_4$ are the coefficients of each activity. By eliminating one activity from the model, the coefficient β_4 for total activity represents the omitted activity component, while the other coefficients represent the consequence of substituting that activity instead of the one removed from the model, while holding other activity types constant.

The partition model was introduced by Willett[91] in the context of epidemiological studies around dietary intake, and an example model can be expressed as follows:

$$y = \beta_1 activity_1 + \beta_2 activity_2 + \beta_3 activity_3 \quad (2.7)$$

The main difference compared to the ISM is the absence of a term for total activities, and therefore, the total time for physical behaviours is not held constant. Each coefficient represents the effect of adding an activity type rather than substituting it.

As well as traditional linear regression, the ISM and partition models consider activity data as absolute values rather than portions of the day.

Compositional isotemporal substitution model

The ISM can be employed within a compositional framework to quantify the change in health associated with reallocations of time between behaviours while respecting the mathematical considerations of compositional data. A method to quantify the change in outcome for changes in physical behaviour has been described in detail by Dumuid et al.[92] and has been used to examine the effect of reallocating various health outcomes and risk factors, including markers of T2D,[93] cardiometabolic risk,[94] and adiposity.[95]

Choosing the right method: Comparing ISM, CoDA, and linear regression

Despite linear regression and ISM being described as inadequate statistical methods for the analysis of movement data since 2014,[96] they are still widely used as analytical tools in time-use epidemiological research. Additionally, despite presenting the same issues as linear regression, ISM yields similar results to CoDA,[97] which complicates the full abandonment of ISM in favour of CoDA. A systematic review[98] comparing the associations between physical behaviours and biomarkers of T2D obtained with ISM, CoDA, and linear regression observed evidence of agreement between ISM and CoDA, but not with linear regression, as also indicated by a recent consensus on methodology to analyse associations with accelerometer-measured PA.[99]

Similar considerations were described by other studies. Dumuid et al.[100] compared associations obtained with ISM and CoDA ISM and found that despite associations being found with both methods, the CoDA ISM produced asymmetrical estimates. That is, a reallocation of one behaviour for another does not predict the exact inverse effect as the reverse reallocation. They also found a non-linear relationship between the amount of time reallocated and body fat percentage, further confirming that traditional linear regression is not an appropriate modelling technique. Biddle et al.[101] found a general agreement between ISM and CoDA ISM, although results obtained using CoDA ISM challenged the notion that replacing sitting with standing is particularly beneficial. Similar to other studies, the authors found asymmetry of reallocations when using CoDA ISM: reducing stepping time produces greater negative effects compared to the benefits of increasing stepping time. The asymmetry of reallocation, as well as the non-linearity of reallocations, are key insights from CoDA studies compared to those using non-CoDA ISM, which also reflects the dose–response relationships observed in experimental studies.[102]

An important limitation of linear regression that is likely to impact the robustness of the estimates is the assumption of a linear relationship between physical behaviours and health. The aforementioned systematic review[103] did not find associations between SB and T2D markers when standard regression was employed, indicating that the assumption of linearity between physical behaviours and outcomes is likely to be wrong. This is further supported by the fact that studies included in this review analysing the participants by quartiles report a dose–response relationship between SB and biomarkers, and independent associations with SB only for very long bouts. For example, Lynch et al.[104] reported significant associations between SB, fasting insulin, and homeostatic model assessment of insulin resistance (HOMA-R) for more than 8.8 hours/day of SB; Healy et al.[105] found a cut-off of 7.24 hours/day of SB that increased fasting insulin; Honda et al.[106] found that more than 10 hours/day of SB significantly increased the odd ratio of insulin resistance and diabetes. This hypothesis has been confirmed by a meta-analysis, which observed a statistically significantly higher risk of death from 9.5 or more hours per day of time spent sedentary.[107]

Moreover, a meta-analysis[108] found evidence of a logarithmic dose–response relationship between SB and all-cause mortality. Findings of a non-linear dose–response association have also been reported for PA in relation to risk of mortality, CVD, and cancer in the general adult population,[109] and in relation to mortality risk in adults with noncommunicable diseases.[110]

There is a general consensus that CoDA, and by extension CoDA ISM, represents the most appropriate method for analysing associations between physical behaviours and health outcomes. By addressing the issue of collinearity inherent in activity data constrained by the 24-hour day, CoDA provides a robust framework for accurate and meaningful interpretations. Aside from methodological considerations around associative analyses, the field still faces challenges in inferring causality within the CoDA framework. While some research has begun to address this issue,[111,112] there is a pressing need for the development of new methods capable of establishing causal relationships in compositional data. Further exploration and refinement of such methodologies will be crucial in advancing the understanding of the impact of physical behaviours on health and informing more effective public health guidelines.

CASE STUDY: UNSUPERVISED CLASSIFICATION OF HUMAN ACTIVITY WITH HIDDEN SEMI MARKOV MODELS

This case study discusses the unsupervised classification of human activity recorded by accelerometers using HSMM. The ability of the HSMM to make correct inferences is assessed by tuning the HSMM parameters to optimise inference performance. Then, the ability of the HSMM approach to be useful for epidemiological studies is investigated, using examples of association between anthropometric data and physical behaviours. In particular, the case study assesses the association between BMI and waist circumference (WC) with SB, MVPA, and LPA, as these associations are associations well documented in the literature.[113,114] This study has been previously published in MDPI Applied System Innovation.[115]

METHODS

The sample used in the analyses was selected from among the participants in the UK Biobank. In the UK Biobank, PA was measured for seven days in a subset of participants between 2013 and 2015 using a wrist-worn triaxial accelerometer.[116] The triaxial acceleration data was captured over a seven-day period at 100 Hz with a dynamic range of ±8 g. The labelled accelerometer data provided by Willetts et al.[117] is used as a reference dataset to validate the HSMM-based method.

An HSMM is fully modelled by the transition probability matrix π observation sequence, y and state duration D whilst the observations are the accelerometer data, either in raw form or as features extracted from the raw data. A diagram illustrating the mathematical model used in this study is shown in Figure 2.2. The system is modelled as a set of states S_i which in this context represents different activities. The system remains in a given state for the duration D_i which is a random variable. Then it transitions to another state, and the whole process forms a Markov chain. The transition parameters $\pi_{i,j}\left(i,j=1,\ldots,n\right)$ form the transition matrix π where $\pi_{i,j}$ is the probability of transition from one state i to another state j. The corresponding observation, or emission variables, denoted by, y form another layer, as shown in Figure 2.2.

An efficient Bayesian inference algorithm has been described by Johnson et al.[118] for the HSMM inference. It is based on the explicit-duration Hierarchical Dirichlet

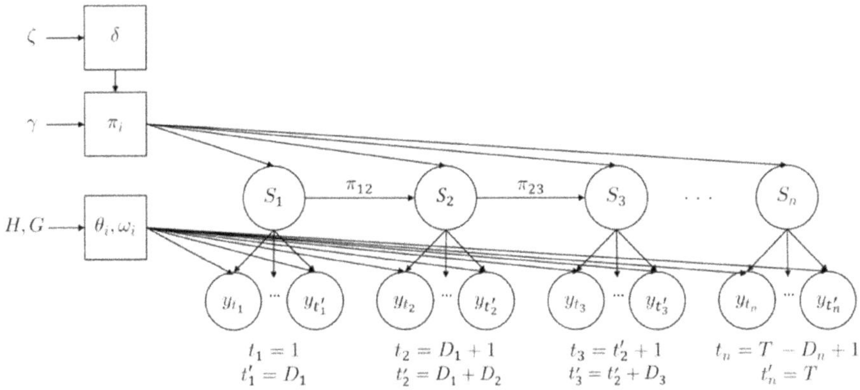

FIGURE 2.2 Diagram of the Hidden Semi-Markov Model. Each state S_i contains a D_i number of observations y The state transitions are modelled by the transition probabilities $\pi_{i,j}$ T denotes the total observation time. The rest of the notation is explained in the main text. Reprinted with permission from Cavallo F. R., Toumazou C., Nikolic K. "Unsupervised Classification of Human Activity with Hidden Semi-Markov Models." *Applied System Innovation*. 2022; 5(4):83. https://doi.org/10.3390/asi5040083

Process HSMM (HDP-HSMM) and sampling algorithms for efficient posterior inference. The HDP-HSMM (ζ, γ, H, G) with n states, parameters ζ, γ, and observation and duration parameter distributions H and G can be summarised as follows:

$$\delta \sim GEM(\zeta) \tag{2.8}$$

$$\pi_i \overset{iid}{\sim} DP(\gamma, \delta) \qquad i = 1, 2, \ldots \tag{2.9}$$

$$(\theta_i, \omega_i) \overset{iid}{\sim} H \times G \tag{2.10}$$

$$D_i \sim g(\omega_i) \tag{2.11}$$

$$y_{t_i:t'_i} \overset{iid}{\sim} h(\theta_i) \tag{2.12}$$

Equation (2.8) represents the sampling of the variable δ using a Dirichlet process with a single parameter ζ—a special case of the Dirichlet process called stick-breaking distribution (GEM).[119] Equation (2.9) denotes the sampling of the transition parameter from a Dirichlet process (DP), with parameters γ and δ (*iid* means independent and identically distributed random variables). Equation (2.10) represents the sampling of the parameters θ and ω from distributions H and G. Then, Equations (2.11) and (2.12) represent the sampling of the state S_i, duration D_i, and observation y, sampled from distributions h and g (which are conjugate distributions of H and G).

The HSMM model parameters are estimated using Bayesian inference. The use of Bayesian inference is ideal in this application, as it allows the incorporation of prior knowledge on observed activity patterns through an appropriate prior distribution. The prior on the state distribution is defined as a Poisson distribution—which models the number of events occurring in a fixed time window—with parameter λ equal to the mean duration of each state. However, for the acceleration values which are observed, a Gaussian prior with mean μ and variance σ^2 is assumed, since HSMM with observations modelled by a Gaussian distribution performs better than other distributions when classifying activities.[120] The parameters are estimated in a Bayesian manner with a Hierarchical Dirichlet Process, which allows for the number of states to be unknown and estimated by the algorithm. The maximum number of states N and the maximum duration D_{max} for each state can be set to reduce training time.

In order to estimate the mean state duration λ the Gamma distribution is used as the conjugate prior for the Poisson distribution. The Gamma distribution has two parameters, α and β, which are related to λ with the relationship $\left(\frac{\alpha}{\beta}=\lambda\right)$. Note that while here the mean duration is referred to as λ, each state will have its own estimated duration λ, such that $\sum_{i=1}^{D}\frac{\lambda_i}{i}=\lambda$. The choice of the hyperparameters α and β is crucial for duration inference. Setting values for α and β determines the effect that the data will have on the posterior: larger values imply larger confidence. Several values for α and β are examined to select the most accurate model. The same approach is adopted for the mean state magnitude μ, and a Gaussian prior with parameters μ_0 and σ_0 is used. The HSMM model parameters were inferred with the Python library *pyhsmm*.[121]

The performance of the model was evaluated by comparing the state inference to the labelled accelerometer data provided by Willetts et al.[122] Firstly, the total weekly time spent in each activity was calculated. Correlation coefficients were calculated for each activity, and the agreement between the two methods was assessed with Bland–Altman analysis. In order to enable such comparisons, the inferred states were mapped to the classified activities (SB, moderate-to-vigorous PA, walking, light tasks, and sleep) based on the Euclidean acceleration.

Additionally, the usefulness of the HSMM-inferred activity for epidemiological research was validated through regression analysis by comparing the coefficients estimated using the HSMM-inferred states and the activities classified by Willetts et al.[123] Two known associations between physical behaviours and anthropometric measures were chosen, namely BMI and WC. Data from the UK Biobank were used for this purpose. The accelerometer data from the UK Biobank were used in the association analyses following a compositional paradigm.

Inference performance and validation results

A sample of 500 participants was selected from the UK Biobank, and a description of the sample is provided in Table 2.1. The average activity duration and acceleration for the labelled data by Willetts et al. were calculated for each activity, namely moderate activity (MPA), sedentary behaviour (SB), sleep, light tasks, and walking (see Table 2.2).

To tune the HSMM, several model parameters were compared. Firstly, the effect of features was assessed by comparing the correlation coefficients of models trained with acceleration magnitude only, with axes acceleration only, with both magnitude and axes acceleration, and with acceleration magnitude and axes angles. For this comparison, the durations prior was set to be of medium informativeness (α = 360 and β = 2). The average duration λ was set to 180 seconds after examining the accelerometer data classified by Willetts et al. Table 2.3 shows the correlation between time spent in activities estimated with the HSMM and with Willett's supervised method for different feature sets. Magnitude alone performed best and was therefore selected for testing of different priors.

The effect of the prior parameters α and β was examined by comparing the medium informative prior (α = 360 and β = 2) with a very informative (α = 1800 and β = 10) and a very non-informative prior (α = 18 and β = 0.1). Table 2.4 shows how different priors affect the correlation between activities classified with the supervised method and HSMM-inferred states. Based on this analysis, the optimal model was the one trained with acceleration magnitude only and with a very

TABLE 2.1
Descriptive characteristics of the sample used in the analyses

Variable	n= 500
Sex (% males)	49
Age (years)	56.08 (7.89)
BMI (kg/m^2)	29.57 (4.22)
Waist circumference (cm)	95.46 (12.47)
MPA (hours/day)	0.79 (0.54)
Walking (hours/day)	4.16 (1.28)
Light tasks (hours/day)	0.63 (0.37)
Sedentary time (hours/day)	8.45 (1.94)
Sleep (hours/day)	9.83 (1.91)
Overall activity (hours/day)	5.58 (1.45)

Reprinted with permission from Cavallo F. R., Toumazou C., Nikolic K. "Unsupervised Classification of Human Activity with Hidden Semi-Markov Models." *Applied System Innovation*. 2022; 5(4):83. https://doi.org/10.3390/asi5040083

Data presented as mean values and standard deviations (in brackets). BMI: body mass index, MPA: moderate physical activity.

TABLE 2.2
Average acceleration and duration of each activity, for the labelled data by Willetts et al.

Activity	Magnitude	X-axis	Y-axis	Z-axis	Duration
MPA	0.084 (0.147)	−0.042 (0.536)	−0.078 (0.433)	0.165 (0.634)	44 s
Sedentary	0.014 (0.022)	−0.077 (0.531)	−0.009 (0.527)	0.14 (0.613)	196 s
Sleep	0.003 (0.007)	−0.015 (0.495)	−0.006 (0.563)	0.114 (0.649)	471 s
Light tasks	0.049 (0.096)	−0.08 (0.446)	−0.065 (0.734)	0.001 (0.411)	53 s
Walking	0.088 (0.078)	−0.027 (0.563)	−0.091 (0.606)	−0.074 (0.401)	125 s
Average	0.048	−0.048	−0.050	0.069	178 s

Reprinted with permission from Cavallo F. R., Toumazou C., Nikolic K. "Unsupervised Classification of Human Activity with Hidden Semi-Markov Models." *Applied System Innovation*. 2022; 5(4):83. https://doi.org/10.3390/asi5040083

Data presented as means and standard deviations. MPA: moderate physical activity.

TABLE 2.3
Effect of different features on the correlation between time spent in activities as estimated by the unsupervised HSMM model and by the supervised model of Willetts et al.

Activity	Magnitude	Magn + Axes	Axes	Magn + Angles
MPA	0.11 [0.02–0.19]	0.12 [0.03–0.21]	0.23 [0.15–0.31]	−0.02 [-0.11–0.07]
Walking	0.5 [0.43–0.57]	0.49 [0.42–0.56]	−0.06 [−0.15–0.03]	0.45 [0.38–0.52]
Light tasks	0.11 [0.03–0.2]	0.1 [0.01–0.18]	0.13 [0.04–0.21]	0.04 [−0.05–0.12]
SB	0.22 [0.13–0.3]	0.26 [0.18–0.34]	−0.04 [−0.13–0.05]	0.18 [−0.05–0.12]
Sleep	0.35 [0.27–0.42]	0.22 [0.14–0.30]	−0.12 [−0.21–0.03]	0.29 [0.21–0.37]
Overall PA	0.59 [0.53–0.64]	0.55 [0.48–0.61]	0.03 [−0.06–0.12]	0.48 [0.41–0.55]
Average (std)	0.31 (0.2)	0.29 (0.19)	0.03 (0.13)	0.09 (0.32)

Reprinted with permission from Cavallo F. R., Toumazou C., Nikolic K. "Unsupervised Classification of Human Activity with Hidden Semi-Markov Models." *Applied System Innovation*. 2022; 5(4):83. https://doi.org/10.3390/asi5040083

MPA: moderate physical activity, SB: sedentary behaviour. Data presented as Pearson's correlation coefficients and 95% confidence intervals, apart from the average which is presented with the standard deviation.

TABLE 2.4
Effect of durations prior on the correlation between time spent in activities as estimated by the HSMM features and by the supervised method of Willetts et al.

Activity	Very informative	Medium	Very uninformative
MPA	0.17 [0.08–0.25]	0.11 [0.02–0.19]	0.18 [0.1–0.2]
Walking	0.35 [0.27–0.43]	0.5 [0.43–0.57]	0.28 [0.2–0.36]
Light tasks	0.19 [0.10–.27]	0.11 [0.03–0.2]	0.19 [0.1–0.27]
Sedentary	0.37 [0.29–0.44]	0.22 [0.13–0.3]	0.48 [0.41–0.55]
Sleep	0.34 [0.26–0.41]	0.35 [0.27–0.42]	0.39 [0.31–0.46]
Overall PA	0.71 [0.67–0.75]	0.59 [0.53–0.64]	0.81 [0.31–0.84]
Average (std)	0.36 (0.19)	0.31 (0.2)	0.39 (0.23)

Reprinted with permission from Cavallo F. R., Toumazou C., Nikolic K. "Unsupervised Classification of Human Activity with Hidden Semi-Markov Models." *Applied System Innovation*. 2022; 5(4):83. https://doi.org/10.3390/asi5040083

Data are presented as Pearson's correlation coefficients and 95% confidence intervals, apart from the average which is presented with the standard deviation. MPA: moderate physical activity.

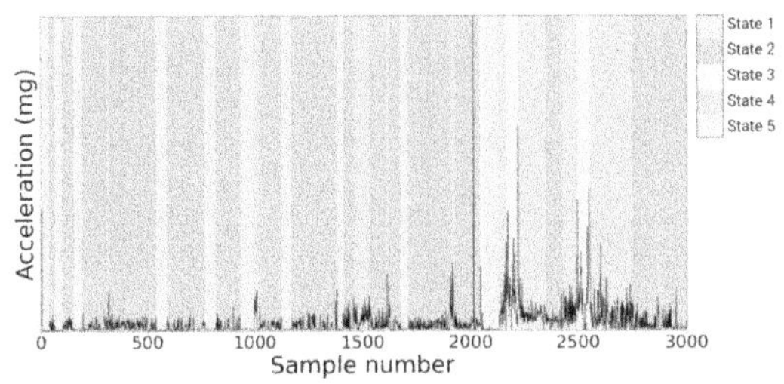

FIGURE 2.3 Example of activity segmentation with HSMM inference on the average acceleration. Reprinted with permission from Cavallo F. R, Toumazou C., Nikolic K. "Unsupervised Classification of Human Activity with Hidden Semi-Markov Models." *Applied System Innovation*. 2022; 5(4):83. https://doi.org/10.3390/asi5040083

uninformative prior. An example of activity segmentation with the chosen model is shown in Figure 2.3.

The model trained with acceleration magnitude only and with a very uninformative prior was chosen. The inferred parameters for the distributions characterising acceleration magnitude and state duration are shown in Table 2.5. The durations found range from 7.5 minutes for sleeping to 8 seconds for walking states, which is reasonable as sleep should be the longest activity and walking is often a transition activity.

TABLE 2.5
Acceleration magnitude and duration for each activity estimated by the HSMM

Activity	Magnitude		Duration (s)
	μ	σ^2	λ
Walking	0.17	0.058	8
Tasks	0.04	0.001	308
SB	0.01	0.0001	366
Sleeping	0.002	$3{\bullet}10^{-6}$	452
MPA	0.11	0.005	254

Reprinted with permission from Cavallo F. R., Toumazou C., Nikolic K. "Unsupervised Classification of Human Activity with Hidden Semi-Markov Models."*Applied System Innovation*. 2022; 5(4):83. https://doi.org/10.3390/asi5040083

MPA: moderate physical activity, SB: sedentary behaviour.

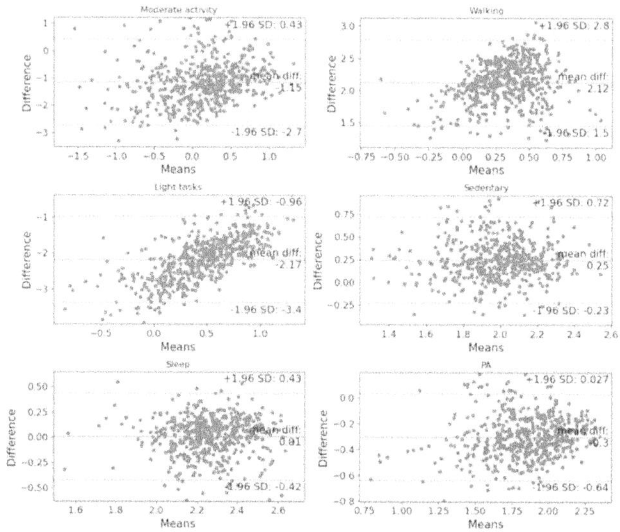

FIGURE 2.4 Bland–Altman plots for method agreement. The methods compared are the classification of activities in the reference dataset by Willetts et al. and the ones inferred by the HSMM. PA denotes the overall physical activity, calculated by summing the time spent in moderate activity, walking, and light tasks. Reprinted with permission from Cavallo F. R., Toumazou C., Nikolic K. "Unsupervised Classification of Human Activity with Hidden Semi-Markov Models." *Applied System Innovation*. 2022; 5(4):83. https://doi.org/10.3390/asi5040083

The Bland–Altman plots in Figure 2.4 show good agreement between the time spent in activities classified by Willetts et al. and the states inferred by the HSMM for sleep, sedentary time, and overall PA (calculated as the sum of walking, moderate activity, and light tasks). On the contrary, there is a visible proportional bias for moderate activity, walking, and light tasks, suggesting poor agreement between the methods for these activities.

Finally, the significance of HSMM-inferred states was tested with linear regression between physical behaviours and anthropometric measures. For this analysis, only the regression coefficients for z_1 are reported, since the regression coefficient β_1 for the first ILR coordinate z_1 represents the strength of the association between the chosen activity and the outcome, while z_{1+i} cannot be interpreted in a meaningful way. The time spent in each activity was calculated for classified activities and HSMM-inferred states and was transformed with an ILR transformation to express it as the time spend in a certain activity with respect to others. The time spent on each activity was calculated for classified activities and HSMM-inferred states and was transformed with an isometric log-ratio transformation to express it as the time spent on a certain activity with respect to others. Given the poor correlation and agreement between moderate activity, light tasks, and walking, only overall PA was considered. From Table 2.6, it can be seen that there is high agreement between the estimated regression coefficients and p-values, indicating that HSMM is a viable method to infer activities from accelerometer data to uncover health-related associations.

Discussion

Lack of PA is the fourth leading cause of global mortality.[124] However, accurately measuring PA is a challenging task, as its objective proxy is body energy expenditure, which can be precisely measured only in laboratory conditions. Various wearable

TABLE 2.6
Regression coefficients for the associations between overall physical activity, sedentary behaviour, sleep and anthropometric measures

Outcome	Activity	Reference	Inferred
BMI	Sedentary	3.27 (<0.0001)	4.51 (<0.0001)
BMI	Physical activity	−3.43 (<0.0001)	−2.74 (0.0006)
BMI	Sleep	0.16 (0.8487)	−1.77 (0.0398)
Waist circumference	Sedentary	9.96 (<0.0001)	12.54 (<0.0001)
Waist circumference	Physical activity	−11.46 (<0.0001)	−7.62 (0.0026)
Waist circumference	Sleep	1.51 (0.5673)	−4.92 (0.0702)

Reprinted with permission from Cavallo F. R., Toumazou C., Nikolic K. "Unsupervised Classification of Human Activity with Hidden Semi-Markov Models." *Applied System Innovation*. 2022; 5(4):83. https://doi.org/10.3390/asi5040083

The coefficients represent the increase in outcome for each extra hour/day spent on each activity. The time spent on physical activity is calculated as the sum of time in moderate activity, walking, and light tasks. The Reference coefficients were estimated by using the labelled dataset by Willetts et al, while the Inferred coefficients were estimated by using the HSMM approach sensory systems have been developed to provide information about PA, but the time-varying signals generated by the sensors need adequate processing in order to produce reliable results. However, sensors and algorithms create results that are strongly dependent on the subject wearing them, and significant research effort has been invested into finding ideal positions for the sensors, such as those for inertial measurements,[125,126] in order to reduce the estimation error. The approach described in this case study employs standard accelerometers that are readily available and an unsupervised algorithm to process the time-varying signal and make inferences about the class of PA.

Unsupervised methods for activity inference are a powerful tool for epidemiological research studies, as they allow trials to be conducted without an ad hoc calibration study for the specific wearable device and population tested, thus saving time and resources. With this aim, this case study presented HSMM as an unsupervised method to classify activity from accelerometer data. A Bayesian framework was used to infer the parameters of the model described in Equations (2.8)–(2.12), and the optimal model was selected based on features used and informativeness of the prior. On the basis of the analysis related to the feature choice and parameter selection, we found the best model to be the one that used the accelerometer magnitude and a very uninformative prior, as it achieved the highest Pearson's correlation between the inferred activities and the reference ones, as shown in Table 2.3 and Table 2.4.

The agreement between the HSMM and supervised classification was evaluated using Bland–Altman plots (Figure 2.4), which showed a good method agreement for SB and overall PA. The model was also validated by reproducing the widely accepted association between PA and sedentary time with BMI and waist circumference using an ILR transformation. The activities inferred by the HSMM produced linear regression coefficients similar to the ones achieved by activities classified using supervised methods (Table 2.6).

A limitation of the current work with the HSMM presented here is the inability to differentiate between activity types, namely moderate activity, walking, and light tasks. In fact, the HSMM seems to only reliably distinguish between activity intensities. This is likely due to the features employed in the inference. While using acceleration magnitude only ensures shorter computational times and easier interpretation of the states, limiting the inference to only this feature may be preventing successful segmentation of different activities with similar intensities. Although the effect of axes angles was explored and did not produce good correlations, future work should explore the use of other time- and frequency-related features.

This case study has been previously published in Cavallo F. R., Toumazou C., Nikolic K. "Unsupervised Classification of Human Activity with Hidden Semi-Markov Models. Applied System Innovation." 2022; 5(4):83. https://doi.org/10.3390/asi5040083

CASE STUDY: ASSOCIATIONS OF TYPE 2 DIABETES CLUSTERS WITH PHYSICAL BEHAVIOURS

As discussed in Chapter 1, several studies have explored the existence of diabetes subtypes through various clustering methods. However, none of the studies in the literature considered the associations between the clusters and lifestyle risk factors. Moreover, no study has examined the interaction between genetics and lifestyle in determining the disease risk associated with the different clusters. While stratification of T2D is a first step towards personalised medicine, it is clear that more work is needed to characterise the roles of lifestyle and genetics on disease progression. Only once this evidence is available, will it be feasible to implement personalised interventions aimed at preventing and managing T2D and its associated comorbidities.

PA is a vital component of lifestyle management interventions, which serve as the first-line therapy for improving glucose control in T2D. Additionally, people with T2D are now also recommended to reduce their sedentary time, as this is associated with increased glucose metabolism biomarkers and several other T2D biomarkers. Several systematic reviews found that biomarkers of glucose metabolism are reduced by substituting SB with both MVPA and LPA. But despite some indications that lower activity intensities can produce benefits, a lack of sufficient evidence prevents the recommendation of specific durations and types of activities to counteract the effects of a sedentary lifestyle.[127]

The aim of this case study is, firstly, to provide further evidence of the combined effects of SB, MVPA, and LPA on T2D biomarkers and common comorbidities, such as Hypercholesterolaemia, hypertension, and CVD. Secondly, to assess whether such effects are different in people belonging to different clusters of T2D, and thus, whether evidence can be found to support personalised lifestyle therapies to manage T2D. In fact, determining the differences in lifestyle behaviours between clusters and their associated risks is essential in developing personalised behavioural change interventions in T2D aimed at lowering the risks associated with poor lifestyle—especially in relation to low PA and high SB. Thirdly, this case study aims to determine whether PA can have a protective effect against a higher genetic risk of developing cardiometabolic conditions. By conducting mediation analyses, this work aims to elucidate the mechanisms through which genetic variants affect cardiometabolic conditions, i.e., directly or through affecting physical behaviours. This knowledge would help understand if physical behaviours are causally related to the risk of CVD, and whether lifestyle therapies can counterbalance the genetic predisposition to cardiometabolic conditions. Additionally, comparing the associations between diabetes clusters can help understand if lifestyle therapies may be more effective for certain types of diabetes compared with others, and thus, predict if an individual would benefit from lifestyle changes.

In summary, the research questions that this case study addresses are:

1. Is there any difference in the association with physical behaviours between the clusters? Is there a difference in the proportional effects of physical behaviours on the risk of developing comorbidities between different clusters?
2. Is the association of genetic variants with comorbidities mediated by physical behaviours? If so, can PA lower the risk of cardiovascular conditions due to unfavourable genetic variants? Do mediations differ between clusters?

To answer the aforementioned research questions, this case study builds on the results presented in the case study included in Chapter 1, which discussed the data-driven clustering of UK Biobank participants with a diagnosis of T2D.

Methods

The sample was selected from among the participants in the UK Biobank, composed of over 500,000 adults aged 37–73 recruited in the UK between 2006 and 2010. At recruitment, participants completed a questionnaire and verbal interview regarding sociodemographic factors, family history, early life exposures, psychosocial factors, environmental factors, lifestyle, and health status. At the same baseline visit, they had physical measurements taken and blood, urine, and saliva samples collected. The study design of the UK Biobank is described in detail by Sudlow et al.[128] Only participants with a self-reported T2D diagnosis, verified by a nurse, were selected and included in the analyses that follow.

PA and SB data were measured through accelerometers for 100,000 participants between 2013 and 2015.[129] The accelerometer data were classified into SB, LPA, or MVPA using the Python *accelerometer* package developed by Willetts et al.[130] Non-wear segments were also included, which Willetts et al. imputed using the average of similar time-of-day data points taken from different days of the week. Only participants with at least three days (72 hours) of data and with data in each one-hour period of the 24-hour cycle, scattered over multiple days, were included. The accelerometer data were used following the compositional paradigm.[131]

Illnesses were self-reported in a questionnaire at the baseline visit and subsequently verified by a nurse through a verbal interview. Three conditions associated with cardiometabolic health were selected: CVD, Hypercholesterolaemia (or high cholesterol), and hypertension. The following diseases were included within CVD: heart/cardiac problem, peripheral vascular disease, venous thromboembolic disease, angina, heart attack/myocardial infarction, heart failure/pulmonary oedema, heart arrhythmia, heart valve problem/heart murmur, cardiomyopathy, pericardial problem, stroke, transient ischaemic attack, subdural haemorrhage/haematoma, subarachnoid haemorrhage, leg claudication/intermittent claudication, arterial embolism, pulmonary embolism, deep venous thrombosis, cerebral aneurysm, myocarditis, atrial fibrillation, rheumatic fever, atrial flutter, Wolff–Parkinson–White syndrome, irregular heartbeat, sick sinus syndrome, supraventricular tachycardia, brain haemorrhage, aortic aneurysm, other venous/lymphatic disease, varicose veins, lymphoedema, ischaemic stroke, mitral valve disease, aortic valve disease, hypertrophic cardiomyopathy, pericarditis, pericardial effusion, aortic aneurysm rupture, aortic dissection, varicose ulcer, mitral valve prolapse, mitral stenosis, mitral regurgitation/incompetence, aortic stenosis, and aortic regurgitation/incompetence.

The genetic markers were selected from a genome-wide association study (GWAS) on the UK Biobank conducted by Doherty et al.[132] who found single nucleotide polymorphisms (SNPs) associated with accelerometer-measured physical behaviours. These SNPs were chosen for mediation analyses as the study shows that physical behaviours share genetic pathways with multiple other traits, including cardiometabolic disease. The SNPs included in the analysis are given in Table 2.7.

TABLE 2.7
SNPs selected for mediation analysis from the GWAS by Doherty et al.

Trait	SNP	Nearest gene	AF	Beta	SE	p-value
MPA	rs10849145	KCNA6	0.825	0.028	0.005	$4.2 \cdot 10^{-9}$
Walking	rs7548269	HP1BP3	0.61	−0.027	0.005	$1.4 \cdot 10^{-8}$
SB	rs26579	MEF2C-AS2	0.415	0.028	0.005	$4.2 \cdot 10^{-9}$
SB	rs25981	EFNA5	0.531	−0.037	0.005	$3.2 \cdot 10^{-12}$
SB	rs1858242	LOC105377146	0.259	−0.028	0.005	$1.9 \cdot 10^{-9}$
SB	rs34858520	CALN1	0.558	−0.051	0.006	$5.8 \cdot 10^{-20}$
SB	rs6870096	NMUR2	0.321	0.090	0.010	$3.1 \cdot 10^{-18}$
SB	rs61776614	FAAP20	0.925	0.043	0.006	$2.7 \cdot 10^{-13}$

AF: allele frequency; MPA: moderate physical activity; SB: sedentary behaviour; SE: standard error.

All the analyses were adjusted for relevant covariates measured at baseline. Self-reported average annual income was used as a measure of socioeconomic status and included in the analyses alongside age, smoking status, alcohol intake frequency, and waist circumference. Self-reported medication use for cholesterol, blood pressure, or insulin was additionally included in regression analyses examining the association between PA and cardiometabolic biomarkers. Sex was not included as a covariate as analyses were stratified by sex to avoid sex-dependent differences in regression models. In fact, women are more susceptible to changes in metabolic risk factors leading to cardiovascular complications in diabetes, and differences in BMI with similar HbA1c between pre-diabetic and diabetic women suggest different relationships between risk factors that are not observed in men.[133]

Linear regression analyses to examine associations with PA were only done on biomarkers reaching medium to high effect sizes (≥0.5)[134] in ANOVA tests. After inspection of residuals and partial regression plots, all biomarkers included as outcomes were transformed with Box–Cox transformations to achieve linearity and homoscedasticity.

Logistic regression was done to assess inter-cluster differences in associations between PA and cardiometabolic conditions (CVD, hypertension, and high cholesterol). All models were adjusted for age, waist circumference, alcohol intake, income, and smoking status. Linear regression models were additionally adjusted for medication use (for cholesterol, blood pressure, and insulin).

Mediation analyses were performed through structural equation modelling with the *lavaan* package[135] in R. Mediation analyses were done both on the whole sample and on the clusters separately. The mediation models were constructed following the diagrams in Figure 2.5, where X represents the genetic variant (i.e., the SNP), $M_1 - M_3$ are the ILR mediators for either SB, LPA, or MPA, Y is the outcome (hypertension, Hypercholesterolaemia, or CVD), C are the covariates, and $a_1 - a_3, b_1 - b_3, c,$ and c' represent the coefficients of the associations between the

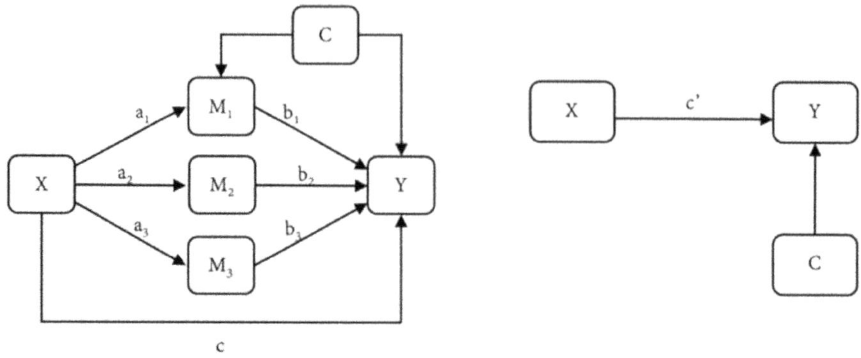

FIGURE 2.5 Mediation models to test the mediating effect of ILR-transformed compositional activity behaviours on the genetic-disease association.

variables they connect. C is also known as the direct effect, $a \cdot b$ as the indirect effect and $c = c + a \cdot b$ as the total effect.

The models were adjusted for age, sex, alcohol intake, smoking, and income. For the models in which Y is the dependent variable, probit regression was performed. To calculate uncertainty measures, the models were bootstrapped with 1,000 iterations.

Results

After excluding participants with missing accelerometer data and covariates, 1,611 observations were included in the analyses. Descriptive statistics of the dataset used in the analyses are presented in Table 2.8 and Table 2.9.

Table 2.10 and Table 2.11 show the differences in physical behaviours and disease incidence between clusters in the regression sample for males and females respectively. For both males and females, Cluster 1 presents lower levels of SBs and higher levels of LPA compared with the other clusters and the sample average. For easier visualisation and comparison to the sample average, Table 2.4 shows the relative average compositional time spent in each activity for each cluster, calculated as the log ratio between the sample-specific and the cluster-specific time spent in each activity. The graphs show that participants in Cluster 1 spend less time on SBs compared to the sample average; they are also the only cluster with more time spent in LPA compared to the sample average. Both Clusters 1 and 2 spend more time than average in MPA. Clusters 3 and 4 are the most sedentary and least active clusters.

Linear regression models examining the association between cardiometabolic markers and PA for each cluster are presented in Table 2.12 and Table 2.13 for males and females, respectively. In Cluster 1, SB was deleteriously associated with apolipoprotein B, while MVPA was beneficially associated with triglycerides for females; for males, MVPA was beneficially associated with HbA1c. For Cluster 2, LPA was deleteriously associated with total cholesterol and LDL-C for females only. For Cluster 3, MVPA was beneficially associated with total cholesterol, glucose, and

TABLE 2.8
Descriptive table of biomarkers for the regression sample and the whole UK Biobank

	Regression sample	UK Biobank
n	1161	Biomarker-dependent
Males %	66	46
Onset age (years)	50.47 (12.56)	N/A
BMI (kg/m^2)	30.81 (5.66)	27.33 (4.77)
Waist circumference (cm)	101.66 (14.14)	90.17 (13.42)
HbA1c (mmol/mol)	51.74 (12.46)	36.14 (6.75)
Glucose (mmol/L)	7.53 (3.37)	5.12 (1.24)
Diastolic BP (mmHg)	81.31 (9.89)	82.46 (10.76)
Systolic BP (mmHg)	142.66 (17.53)	138.75 (19.35)
Pulse rate (bpm)	73 (12.76)	68.87 (11.65)
Apolipoprotein A (g/L)	1.41 (0.25)	1.54 (0.27)
Apolipoprotein B (g/L)	0.83 (0.2)	1.03 (0.24)
Lipoprotein A (nmol/L)	41.97 (48.18)	44.64 (49.21)
Triglycerides (mmol/L)	2.03 (1.15)	1.75 (1.02)
Cholesterol (mmol/L)	4.45 (0.96)	5.69 (1.15)
LDL direct (mmol/L)	1.22 (0.33)	3.56 (0.87)
HDL-C (mmol/L)	2.66 (0.72)	1.45 (0.38)
CRP	2.94 (4.39)	2.59 (4.35)

Data presented as means and standard deviation.

TABLE 2.9
Descriptive table of lifestyle variables for the regression sample

Variable	Mean (SD)
SB (h/day)	8.68 (1.79)
LPA (min/day)	214.66 (87.55)
MVPA (min/day)	47.93 (38.74)
CVD (%)	21
Hypertension (%)	38
High cholesterol (%)	37
Alcohol intake frequency (%)	
Daily or almost daily	18
Three or four times a week	20
Once or twice a week	25
One to three times a month	13
Special occasions only	10
Current or past smokers (%)	56
Cholesterol medication users (%)	77
Hypertension medication users (%)	61

Data presented as means and standard deviation, or percentages for categorical variables.

TABLE 2.10
Characteristics of the regression sample (males only), split by cluster

	Cluster 1 (n = 332)	Cluster 2 (n = 259)	Cluster 3 (n = 272)	Cluster 4 (n = 200)
Age (years)	59.06 (7.57)	60.61 (6.58)	59.58 (6.55)	59.32 (6.24)
WC (cm)	98.59 (12.71)	103.51 (12.58)	107.28 (12.31)	109.34 (11.61)
SB (h/day)	8.43 (1.68)	8.99 (1.77)	9.07 (1.84)	9.05 (1.78)
LPA (min/day)	228.21 (85.72)	206.55 (83.52)	205.96 (82.99)	209.5 (79.18)
MVPA (min/day)	42.0 (31.61)	44.68 (33.42)	38.33 (29.15)	39.0 (33.1)
CVD (%)	23	25	23	22
Hypertension (%)	62	62	66	62
High cholesterol (%)	35	37	39	40
Alcohol intake (%)				
Daily or almost daily	23	19	21	25
3–4 times a week	24	25	22	18
1–2 times a week	25	26	27	23
1–3 times a month	10	11	13	13
Special occasions only	18	19	17	21
Current/past smokers (%)	57	61	62	66
Medication users (%)				
Cholesterol	74	81	79	79
Hypertension	57	65	73	68
Insulin	28	13	13	16

HbA1c in females, while it was only associated with HbA1c in males. Additionally, for Cluster 3 in males only, SB was negatively associated with apolipoprotein A and HDL-C, while LPA was associated with total cholesterol. Finally, Cluster 4 had significant associations for males only, with MVPA being beneficially associated with glucose and HbA1c. In the sex-combined analyses, apolipoprotein A was significantly associated with LPA in Cluster 3, and SB in Cluster 2. Apolipoprotein B was significantly associated with SB in Clusters 2 and 4. MVPA was associated with lower glucose in Cluster 4 only. In Cluster 4, HbA1c was associated with LPA with a large effect size (–3.2). HDL-C was associated with LPA in Clusters 2 and 3, while LDL-C was associated with LPA in Cluster 3 and SB in Cluster 4. Triglycerides were only associated with MVPA in Cluster 1.

Associations between PA and comorbidities of T2D are presented in Table 2.12 and Table 2.13. In Cluster 1, LPA is associated with significantly reduced odds of hypertension in both males and females, and for males, it is additionally associated with improved odds of CVD. In Cluster 1 for males only, SB is associated with increased odds of CVD, and MVPA is associated with increased odds of high cholesterol. Cluster 2 had only one significant association in males, with SB associated with increased odds of hypertension. In Cluster 3, for females only, MVPA led to reduced odds of CVD, while LPA decreased and SB increased the odds of hypertension.

TABLE 2.11
Characteristics of the regression sample (females only), split by cluster

	Cluster 1 (n = 196)	Cluster 2 (n = 148)	Cluster 3 (n = 119)	Cluster 4 (n = 85)
Age (years)	57.57 (7.72)	58.28 (6.37)	59.29 (6.29)	57.15 (7.14)
WC (cm)	89.29 (14.36)	98.33 (12.97)	101.97 (13.93)	105.93 (13.0)
SB (h/day)	8.1 (1.73)	8.33 (1.75)	8.48 (1.62)	8.85 (1.9)
LPA (min/day)	231.07 (96.14)	213.56 (94.87)	213.86 (92.45)	191.67 (88.5)
MVPA (min/day)	67.5 (52.63)	58.79 (46.47)	59.76 (42.37)	51.98 (36.83)
CVD (%)	11	19	14	15
Hypertension (%)	54	65	65	61
High cholesterol (%)	36	35	33	42
Alcohol intake (%)				
Daily or almost daily	13	9	6	6
3–4 times a week	17	9	7	15
1–2 times a week	19	26	22	19
1–3 times a month	12	12	20	15
Special occasions only	39	44	45	45
Current/past smokers (%)	42	46	44	52
Medication users (%)				
Cholesterol	71	76	78	75
Hypertension	45	61	52	62
Insulin	44	16	12	15

In the sex-combined analyses shown in Table 2.14, CVD odds were significantly reduced with increasing LPA in Cluster 1 with a large effect size (0.4). Additionally, in Cluster 1, MVPA was associated with lower odds of high cholesterol, while LPA was associated with lower odds of hypertension. LPA was also associated with lower odds of hypertension in Cluster 3.

From the assessment of the proportion of day spent in SB, LIPA, and MVPA for each cluster (see Figure 2.6), it can be seen that Cluster 1 spends more time in LIPA and MVPA compared to the sample average. Individuals belonging to this cluster also spent less time in SB. Conversely, Cluster 2 spent slightly more time in SB and less time in LIPA compared to the average, but still engaged in more MVPA. Clusters 3 and 4, on the other hand, spent a higher proportion of their day being sedentary, and less time engaging in physical activity, of both low and moderate-to-vigorous intensity.

In whole sample analyses, SB was found to be a mediator of the relationship between the SNPs rs25981 and hypertension, and between rs68700 and hypertension and CVD. The negative effect that rs25981 has on hypertension seems to be mainly due to its effect on SB. That is, rs25981 does not have a direct effect on hypertension, but it is associated with the odds of hypertension through the inverse association

TABLE 2.12
Compositional models for cardiometabolic markers and comorbidities in male clusters

Variable	Cluster 1 (n = 332)	Cluster 2 (n = 259)	Cluster 3 (n = 272)	Cluster 4 (n = 200)
Apolipoprotein A				
MVPA	0.006 (0.64)	0.005 (0.64)	0.012 (0.25)	0.001 (0.96)
LPA	0.007 (0.72)	0.023 (0.21)	0.03 (0.07)	0.032 (0.16)
SB	−0.031 (0.3)	−0.014 (0.62)	−0.006 (0.8)	−0.032 (0.31)
Apolipoprotein B				
MVPA	0.002 (0.92)	−0.013 (0.44)	−0.013 (0.45)	−0.023 (0.25)
LPA	0.013 (0.65)	0.038 (0.19)	0.047 (0.1)	−0.059 (0.12)
SB	−0.025 (0.55)	−0.035 (0.43)	0.03 (0.49)	−0.06 (0.26)
Cholesterol				
MVPA	−0.018 (0.22)	−0.01 (0.49)	−0.018 (0.18)	−0.027 (0.08)
LPA	0.014 (0.54)	0.036 (0.15)	**0.047 (0.03)** *	−0.012 (0.67)
SB	0.014 (0.7)	−0.042 (0.26)	0.001 (0.97)	−0.051 (0.2)
Glucose				
MVPA	−025 (0.1)	−008 (0.54)	−0.009 (0.44)	−0.035 (0.01)
LPA	0.03 (0.23)	−0.006 (0.79)	−0.003 (0.87)	0.006 (0.8)
SB	0.055 (0.14)	−0.03 (0.36)	−0.026 (0.39)	−0.006 (0.86)
HbA1c				
MVPA	**−0.02 (0.03)** *	−0.001 (0.9)	**−0.017 (0.05)** *	**−0.025 (0.01)** *
LPA	0.018 (0.21)	0.016 (0.32)	0.005 (0.69)	0.004 (0.82)
SB	0.025 (0.24)	−0.027 (0.25)	−0.007 (0.75)	0.002 (0.92)
HDL-C				
MVPA	0.008 (0.65)	0.005 (0.76)	0.017 (0.23)	0.009 (0.57)
LPA	0.005 (0.85)	0.036 (0.17)	0.041 (0.07)	0.012 (0.71)
SB	−0.038 (0.35)	−0.005 (0.9)	0.004 (0.9)	−0.057 (0.19)
LDL-C				
MVPA	−0.016 (0.39)	−0.009 (0.62)	−0.022 (0.21)	−0.03 (0.11)
LPA	0.012 (0.69)	0.035 (0.28)	**0.061 (0.04)** *	−0.034 (0.33)
SB	0.008 (0.85)	−0.045 (0.35)	0.005 (0.91)	−-0.078 (0.11)
Triglycerides				
MVPA	−0.008 (0.68)	−0.003 (0.8)	−0.009 (0.43)	−0.034 (0.15)
LPA	0.009 (0.79)	−0.003 (0.85)	0.002 (0.91)	0.017 (0.7)
SB	−0.02 (0.69)	−0.002 (0.95)	−0.02 (0.46)	0.042 (0.49)
CVD				
MVPA	1.097 (0.63)	0.709 (0.07)	0.965 (0.85)	1.01 (0.96)
LPA	**0.3 (0.0001)** *	0.793 (0.48)	0.778 (0.41)	0.737 (0.46)
SB	1.709 (0.26)	1.363 (0.55)	1.288 (0.6)	0.819 (0.74)

(Continued)

TABLE 2.12 (CONTINUED)
Compositional models for cardiometabolic markers and comorbidities in male clusters

Variable	Cluster 1 (n = 332)	Cluster 2 (n = 259)	Cluster 3 (n = 272)	Cluster 4 (n = 200)
High cholesterol				
MVPA	**1.477 (0.02)** *	0.942 (0.74)	1.001 (1)	0.914 (0.63)
LPA	0.764 (0.31)	0.95 (0.87)	0.97 (0.91)	1.428 (0.32)
SB	0.575 (0.18)	1.461 (0.41)	1.899 (0.13)	1.272 (0.62)
Hypertension				
MVPA	1.099 (0.57)	1.078 (0.67)	0.816 (0.26)	0.89 (0.54)
LPA	0.599 (0.06)	1.154 (0.64)	**0.529 (0.05)**	1.575 (0.2)
SB	1.954 (0.1)	0.635 (0.33)	1.269 (0.59)	1.208 (0.7)

Statistically significant associations ($p < 0.05$) are in bold and marked with *. The models were adjusted for age, income, smoking status, alcohol consumption, and use of medication for high cholesterol, high blood pressure, and insulin. The regression coefficients for the model correspond to a change in the log ratio of the given behaviour to the others. Data are presented as coefficients and *p*-values. Biomarkers have been transformed with Box–Cox transformation

Each biomarker was regressed against the proportion of the day spent on each behaviour: sedentary behaviour (SB), light physical activity (LPA), and moderate-to-vigorous physical activity (MVPA)

TABLE 2.13
Compositional models for cardiometabolic markers and comorbidities in female clusters

Variable	Cluster 1 (n = 196)	Cluster 2 (n = 148)	Cluster 3 (n = 119)	Cluster 4 (n = 85)
Apolipoprotein A				
MVPA	–0.003 (0.79)	0.003 (0.82)	0.002 (0.92)	0.013 (0.5)
LPA	0.004 (0.84)	0.027 (0.21)	0.029 (0.22)	–0.016 (0.56)
SB	0.047 (0.18)	–0.057 (0.11)	–0.044 (0.32)	0.014 (0.77)
Apolipoprotein B				
MVPA	–0.015 (0.46)	–0.024 (0.31)	–0.005 (0.86)	0.036 (0.24)
LPA	–0.057 (0.14)	0.03 (0.42)	–0.011 (0.77)	–0.027 (0.56)
SB	0.081 (0.18)	–0.067 (0.27)	–0.026 (0.71)	–0.167 (0.03)
Cholesterol				
MVPA	–0.013 (0.38)	–0.02 (0.27)	**–0.053 (0.01)** *	0.022 (0.41)
LPA	–0.01 (0.7)	0.047 (0.11)	0.033 (0.27)	–0.004 (0.92)
SB	0.042 (0.33)	**–0.111 (0.02)** *	–0.004 (0.95)	**–0.141 (0.04)** *
Glucose				
MVPA	–0.002 (0.88)	–0.014 (0.34)	**–0.038 (0.04)** *	–0.018 (0.36)
LPA	0.026 (0.33)	–0.001 (0.95)	–0.008 (0.74)	–0.023 (0.44)

(Continued)

TABLE 2.13 (CONTINUED)
Compositional models for cardiometabolic markers and comorbidities in female clusters

Variable	Cluster 1 (n = 196)	Cluster 2 (n = 148)	Cluster 3 (n = 119)	Cluster 4 (n = 85)
SB	−0.057 (0.18)	0.004 (0.91)	0.028 (0.56)	−0.058 (0.24)
HbA1c				
MVPA	−0.004 (0.74)	−0.012 (0.34)	**−0.054 (0.0004)** *	−0.013 (0.43)
LPA	0.023 (0.28)	0.006 (0.76)	0.032 (0.13)	−0.003 (0.9)
SB	−0.028 (0.41)	**−0.065 (0.05)** *	0.028 (0.49)	−0.033 (0.42)
HDL-C				
MVPA	0.004 (0.83)	−0.006 (0.75)	−0.008 (0.71)	0.032 (0.22)
LPA	0.003 (0.92)	0.042 (0.17)	0.022 (0.49)	−0.046 (0.24)
SB	0.055 (0.28)	−0.045 (0.37)	−0.029 (0.63)	−0.002 (0.97)
LDL-C				
MVPA	−0.021 (0.21)	−0.02 (0.36)	−0.045 (0.08)	0.039 (0.19)
LPA	−0.036 (0.25)	0.052 (0.15)	0.02 (0.59)	−0.01 (0.82)
SB	0.071 (0.15)	−0.108 (0.06)	0.011 (0.88)	**−0.191 (0.01)** *
Triglycerides				
MVPA	**−0.06 (0.02)** *	0.011 (0.39)	−0.028 (0.1)	−0.01 (0.81)
LPA	−0.033 (0.5)	−0.009 (0.68)	0.01 (0.69)	−0.001 (0.99)
SB	0.128 (0.1)	−0.007 (0.83)	0.052 (0.26)	0.05 (0.62)
CVD				
MVPA	1.408 (0.29)	1.713 (0.1)	**0.439 (0.03)** *	1.434 (0.44)
LPA	1.037 (0.95)	0.622 (0.33)	0.727 (0.53)	1.801 (0.43)
SB	1.318 (0.76)	1.213 (0.8)	0.631 (0.66)	0.304 (0.3)
High cholesterol				
MVPA	1.32 (0.16)	1.23 (0.42)	0.76 (0.34)	0.692 (0.26)
LPA	0.769 (0.47)	1.182 (0.67)	1.07 (0.87)	2.328 (0.1)
SB	1.161 (0.8)	0.498 (0.26)	2.896 (0.18)	1.544 (0.6)
Hypertension				
MVPA	1.132 (0.52)	1.359 (0.2)	0.669 (0.17)	0.725 (0.34)
LPA	**0.446 (0.03)** *	0.753 (0.46)	**0.319 (0.02)** *	0.566 (0.27)
SB	1.042 (0.94)	0.613 (0.43)	3.156 (0.15)	0.824 (0.82)

Statistically significant associations ($p < 0.05$) are in bold and marked with *. The models were adjusted for age, income, smoking status, alcohol consumption, and use of medication for high cholesterol, high blood pressure, and insulin. The regression coefficients for the model correspond to a change in the log ratio of the given behaviour to the others. Data are presented as coefficients and *p*-values. Biomarkers have been transformed with Box–Cox transformation.

Each biomarker was regressed against the proportion of the day spent in each behaviour: sedentary behaviour (SB), light physical activity (LPA) and moderate-to-vigorous physical activity (MVPA)

TABLE 2.14
Compositional models for cardiometabolic markers and comorbidities in sex-combined clusters

Variable	Cluster 1 (n = 528)	Cluster 2 (n = 407)	Cluster 3 (n = 285)	Cluster 4 (n = 391)
Apolipoprotein A				
MVPA	−0.013 (0.44)	0.004 (0.82)	−0.004 (0.81)	−0.011 (0.52)
LPA	0.017 (0.55)	0.047 (0.07)	**0.062 (0.01)** *	0.026 (0.39)
SB	−0.005 (0.9)	**−0.092 (0.02)** *	−0.029 (0.45)	−0.034 (0.43)
Apolipoprotein B				
MVPA	−0.011 (0.33)	−0.022 (0.08)	−0.023 (0.09)	−0.018 (0.28)
LPA	−−0.001 (0.95)	0.04 (0.06)	0.033 (0.11)	−0.045 (0.13)
SB	0.015 (0.59)	**−0.063 (0.05)** *	0.017 (0.63)	**−0.085 (0.05)** *
Cholesterol				
MVPA	−0.06 (0.26)	−0.096 (0.1)	−0.094 (0.13)	−0.027 (0.75)
LPA	−0.01 (0.91)	**0.23 (0.02)** *	**0.229 (0.02)** *	−0.08 (0.59)
SB	0.122 (0.38)	**−0.314 (0.04)** *	0.03 (0.85)	**−0.482 (0.02)** *
Glucose				
MVPA	−0.152 (0.48)	−0.18 (0.34)	0.059 (0.78)	**−0.539 (0.03)** *
LPA	0.053 (0.88)	−0.063 (0.84)	−0.526 (0.11)	−0.589 (0.19)
SB	0.458 (0.42)	−0.059 (0.9)	−0.303 (0.57)	−0.391 (0.55)
HbA1c				
MVPA	−0.638 (0.37)	−0.032 (0.97)	−0.29 (0.73)	−1.776 (0.06)
LPA	1.741 (0.14)	1.011 (0.44)	−1.418 (0.27)	**−3.216 (0.05)** *
SB	1.29 (0.49)	−2.348 (0.25)	1.249 (0.56)	1.676 (0.49)
HDL-C				
MVPA	−0.01 (0.67)	−0.006 (0.74)	−0.009 (0.54)	0.0001 (1)
LPA	0.017 (0.64)	**0.062 (0.03)** *	**0.055 (0.02)** *	−0.001 (0.99)
SB	−0.005 (0.93)	−0.079 (0.08)	−0.001 (0.98)	−0.049 (0.24)
LDL-C				
MVPA	−0.043 (0.24)	−0.068 (0.11)	−0.076 (0.1)	−0.021 (0.71)
LPA	−0.016 (0.8)	0.145 (0.04)	**0.139 (0.05)** *	−0.096 (0.33)
SB	0.09 (0.36)	−0.196 (0.08)	0.05 (0.67)	**−0.368 (0.01)** *
Triglycerides				
MVPA	**−0.034 (0.01)** *	0.002 (0.89)	−0.027 (0.22)	−0.073 (0.4)
LPA	0.011 (0.64)	−0.003 (0.91)	0.013 (0.7)	0.039 (0.8)
SB	0.026 (0.47)	−0.027 (0.47)	−0.011 (0.85)	0.206 (0.36)
CVD				
MVPA	1.109 (0.51)	0.929 (0.64)	0.803 (0.19)	1.082 (0.68)
LPA	**0.396 (0.0002)** *	0.643 (0.09)	0.779 (0.33)	0.935 (0.85)
SB	1.534 (0.29)	1.482 (0.34)	1.254 (0.6)	0.671 (0.44)

(Continued)

TABLE 2.14 (CONTINUED)
Compositional models for cardiometabolic markers and comorbidities in sex-combined clusters

Variable	Cluster 1 (n = 528)	Cluster 2 (n = 407)	Cluster 3 (n = 285)	Cluster 4 (n = 391)
High cholesterol				
MVPA	**1.388 (0.01)** *	1.025 (0.86)	0.934 (0.63)	0.827 (0.23)
LPA	0.772 (0.21)	1.013 (0.96)	1.056 (0.81)	1.692 (0.07)
SB	0.714 (0.31)	0.923 (0.82)	2.038 (0.06)	1.299 (0.52)
Hypertension				
MVPA	1.12 (0.36)	1.153 (0.31)	0.775 (0.09)	0.845 (0.3)
LPA	**0.563 (0.01)** *	0.988 (0.96)	**0.439 (0.0016)** *	1.159 (0.6)
SB	1.609 (0.14)	0.684 (0.3)	1.615 (0.22)	1.189 (0.68)

Statistically significant associations ($p < 0.05$) are in bold and marked with *. The models were adjusted for age, income, smoking status, alcohol consumption, use of medication for high cholesterol, high blood pressure, and insulin. The regression coefficients for the model correspond to a change in the log-ratio of the given behaviour to the others. Data are presented as coefficients and p-values. Biomarkers have been transformed with Box–Cox transformation.

Each biomarker or comorbidity was regressed against the proportion of the day spent in each behaviour: sedentary behaviour (SB), light phys-ical activity (LPA) and moderate-to-vigorous physical activity (MVPA)

with SB. Similarly, rs6870096 also acts on the risk of hypertension through its effect on SB, by increasing the time spent in SB.

The relationships between hypertension and the two variants that were significant in sample-wide analysis were examined in cluster-specific analyses to explore potential differences between clusters. The results were non-significant, apart from one significant mediation effect between rs6870096, SB and CVD in Cluster 1 only. The significant results from the mediation analyses are shown in Figure 2.7.

Discussion

In Chapter 1, four phenotypically distinct clusters of T2D were found based on 16 cardiometabolic markers routinely measured in clinical settings. The clustering highlighted the presence of different types of diabetes, characterised by different cardiometabolic profiles. Cluster 1 was characterised by a better lipidic profile compared to other clusters, and an increased insulin use suggests that this kind of diabetes subtype may not be caused by an unhealthy lifestyle, but rather by impaired mechanisms of glucose control, more similar to type 1 diabetes. Clusters 3 and 4 presented higher onset age, obesity, and higher levels of triglycerides and lipids. Paired with the higher incidence of CVD, hypertension, and hyperglycaemia, the clusters may be representative of a diabetes subtype caused by unhealthy diets and inactivity. Levels of physical behaviours further support this hypothesis: Cluster 1

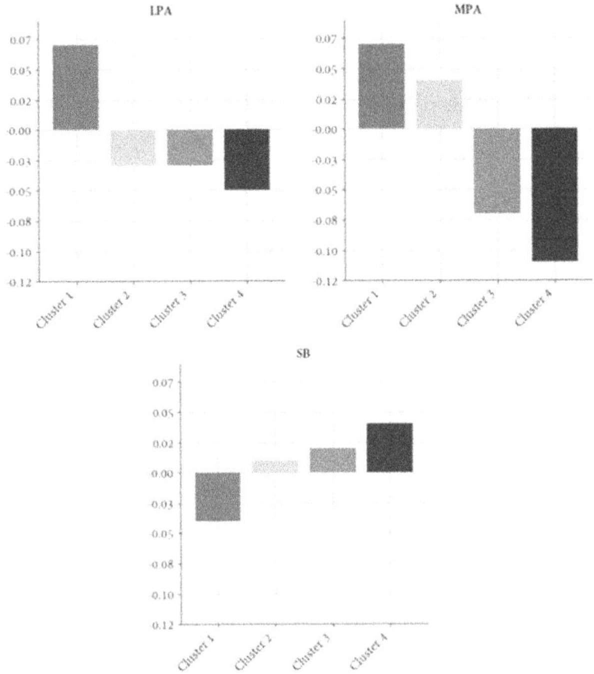

FIGURE 2.6 Compositional importance of the mean daily time spent in physical behaviours for each cluster. The left axis gives the logratio value between the group mean and the total sample mean. SB is sedentary behaviour, MPA is moderate physical activity and LPA is light physical activity.

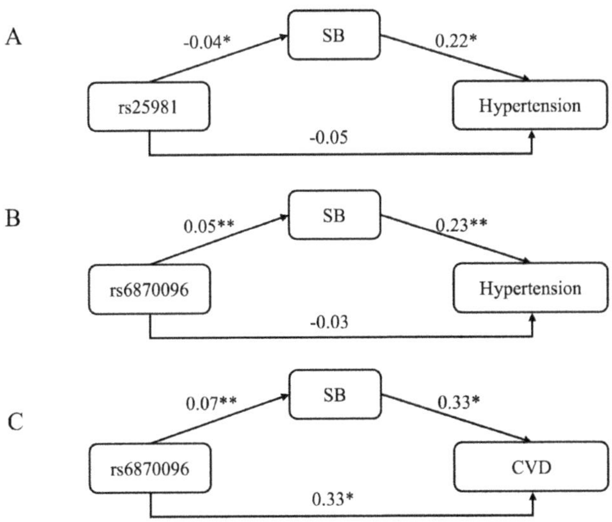

FIGURE 2.7 Significant mediation results. Significant mediation results for whole sample analysis (A, B) and for Cluster 1 (C).

spend less time sedentary compared to the sample average and more time engaging in PA compared to the sample average. Conversely, Clusters 3 and 4 are the most sedentary and least active clusters. Cluster 2 has a cardiometabolic profile between Clusters 1 and 3, is sedentary, but also active.

The clusters presented different strengths of associations between biomarkers and comorbidities of T2D, and physical behaviours. In particular, in Cluster 1, LPA was significantly associated with reduced odds of CVD and hypertension, while MVPA was associated with reduced odds of high cholesterol. In Cluster 3, higher LPA was associated with lower hypertension, while the association between high cholesterol and SB almost reached significance, indicating that reducing SB may be beneficial for reducing the risk of high cholesterol in Cluster 3. Finally, there was a significant association between HbA1c and LPA in Cluster 4, with a coefficient of −3.2—much larger than in the other clusters. This suggests that increasing LPA may be effective in reducing HbA1c in Cluster 4, but further research should explore possible causal mechanisms, and verify this hypothesis through randomised controlled trials.

In whole-sample analyses, SB was a significant mediator of the relationship between the SNPs rs25981 and hypertension, and between rs68700 and hypertension and CVD. However, this result was not replicated in cluster-specific analyses due to non-significance, and thus, it is not possible to comment on whether differences exist between clusters. Only one significant mediation was significant between rs6870096, SB and CVD in Cluster 1. However, due to the other clusters having smaller sample sizes, it is not possible to compare the differences between clusters as the significance of the mediation for Cluster 1 may be due to the other analyses being underpowered. For other genetic variants and physical behaviours, it is possible that there is either no relationship or no mediation. The former (no relationship) may be due to the use of ILR-transformed physical behaviours in these analyses. In fact, Doherty et al. found the SNPs used in this work using non-ILR transformed accelerometer data, which may have led to different associations between genetic variants and activity. The latter (no mediation) suggests that some activity-related traits affect disease directly when ILR accelerometer data is used, and that there may be a genetic connection between physical behaviour and cardiometabolic diseases, as also suggested by Doherty et al.[136]

While the results presented above suggest some differences in associations between clusters, it is legitimate to wonder if such differences are due to different sample sizes and, therefore, to different power levels achieved by the statistical models. A post-hoc power analysis could be performed to calculate the power achieved in each statistical analysis; however, using the estimated effect sizes to calculate post-hoc power would yield extremely noisy measures of power, especially if using a non-statistically significant estimate.

However, post-hoc power analysis is not necessary, as the ultimate aim of statistical analyses on clinical datasets is to find clinically, rather than statistically, significant associations. Looking at Table 2.12, Table 2.13, and Table 2.14 only higher regression coefficients are statistically significant and, thus, more likely to have practical implications. Lower coefficients, despite not achieving statistical significance, can still be used to compare differences between the clusters. For example, Table 2.14 shows that people in Cluster 1 have significantly lower odds of CVD for increased LPA; the other clusters do not present significant associations, but even if they did, the effect of LPA on CVD risk would be smaller than for Cluster 1. Thus, a practical implication may be that the cluster of T2D can determine how much LPA is required to observe the same reduction in CVD risk as for Cluster 1.

In conclusion, the study underscores the importance of recognising the heterogeneity within T2D for developing targeted and effective personalised interventions. Understanding the distinct profiles and associations within each cluster can inform more precise and tailored therapeutic strategies, ultimately improving patient outcomes.

ABBREVIATIONS

PA	Physical activity
LPA	Low intensity physical activity
MPA	Moderate intensity physical activity
MVPA	Moderate-to-vigorous intensity physical activity
SB	Sedentary behaviour
MET	Metabolic equivalent of task
HMM	Hidden Markov Model
HOMA-IR	Homeostatic model assessment of insulin resistance
GWAS	Genome-wide association study
HSMM	Hidden Semi-Markov Model
T2D	Type 2 diabetes
CVD	Cardiovascular disease
CoDA	Compositional data analysis
ILR	Isometric log ratio
ISM	Isotemporal substitution model

NOTES

1. M. J. Davies et al., "Management of Hyperglycemia in Type 2 Diabetes, 2018: A Consensus Report by the American Diabetes Association (ADA) and the European Association for the Study of Diabetes (EASD)," *Diabetes Care* 41, no. 12 (2018): 2669–701.
2. M. S. Tremblay et al., "Sedentary Behavior Research Network (SBRN) – Terminology Consensus Project Process and Outcome," *Int J Behav Nutr Phys Act* 14, no. 1 (2017): 75.
3. K. Wirth et al., "Biomarkers Associated with Sedentary Behaviour in Older Adults: A Systematic Review," *Ageing Res Rev* 35 (2017): 87–111.
4. L. A. Brocklebank et al., "Accelerometer-Measured Sedentary Time and Cardiometabolic Biomarkers: A Systematic Review," 76 (2015): 92–102.
5. Department of Health and Social Care et al., "UK Chief Medical Officers' Physical Activity Guidelines," (2019), https://assets.publishing.service.gov.uk/government/uploads/system/uploads/attachment_data/file/832868/uk-chief-medical-officers-physical-activity-guidelines.pdf (Accessed 20/09/2024).
6. Australian Government Department of Health, "Australia's Physical Activity and Sedentary Behaviour Guidelines. Fact Sheet: Adults (18–64 Years)," (2021), https://www.health.gov.au/resources/publications/physical-activity-and-sedentary-behaviour-guidelines-adults-18-to-64-years-fact-sheet (Accessed 29/09/2024).

7. P. C. Dempsey et al., "New Global Guidelines on Sedentary Behaviour and Health for Adults: Broadening the Behavioural Targets," *Int J Behav Nutr Phys Act* 17, no. 1 (2020): 1–12.
8. F. C. Bull et al., "World Health Organisation 2020 Guidelines on Physical Activity and Sedentary Behaviour," *Br J Sports Med* 54, no. 24 (2020): 1451–62.
9. R. Ross et al., "Canadian 24-Hour Movement Guidelines for Adults Aged 18–64 Years and Adults Aged 65 Years or Older: An Integration of Physical Activity, Sedentary Behaviour, and Sleep," *Appl Physiol Nutr Metab* 45, no. 10 (Suppl. 2) (2020): S57–S102.
10. A. D. Okely et al., "A Collaborative Approach to Adopting/Adapting Guidelines – The Australian 24-Hour Movement Guidelines for the Early Years (Birth to 5 Years): An Integration of Physical Activity, Sedentary Behavior, and Sleep," *BMC Public Health* 17, no. 5 (2017): 869.
11. New Zealand Ministry of Health, "Sit Less, Move More, Sleep Well: Active Play Guidelines for under-Fives," Wellington: Ministry of Health (2017), https://www.health.govt.nz/publications/sit-less-move-more-sleep-well-active-play-guidelines-for-under-fives (Accessed 29/09/2024).
12. C. E. Draper et al., "The South African 24-Hour Movement Guidelines for Birth to 5 Years: An Integration of Physical Activity, Sitting Behavior, Screen Time, and Sleep," *J Phys Act Health* 17, no. 1 (2020): 109–19.
13. World Health Organization, "Guidelines on Physical Activity, Sedentary Behaviour and Sleep for Children under 5 Years of Age," World Health Organization, Geneva (2019), https://www.who.int/publications/i/item/9789241550536
14. S. F. M. Chastin et al., "How Does Light-Intensity Physical Activity Associate with Adult Cardiometabolic Health and Mortality? Systematic Review with Meta-Analysis of Experimental and Observational Studies," 53 (2019): 370–76.
15. D. P. Bailey and C. D. Locke, "Breaking up Prolonged Sitting with Light-Intensity Walking Improves Postprandial Glycemia, but Breaking up Sitting with Standing Does Not," *J Sci Med Sport* 18, no. 3 (2015): 294–98.
16. S. A. Newsom et al., "A Single Session of Low-Intensity Exercise Is Sufficient to Enhance Insulin Sensitivity into the Next Day in Obese Adults," *Diabetes Care* 36, no. 9 (2013): 2516–22.
17. R. J. F. Manders, J. W. M. Van Dijk, and L. J. C. Van Loon, "Low-Intensity Exercise Reduces the Prevalence of Hyperglycemia in Type 2 Diabetes," *Med Sci Sports Exerc* 42, no. 2 (2010): 219–25.
18. Chastin et al., "How Does Light-Intensity Physical Activity Associate with Adult Cardiometabolic Health and Mortality? Systematic Review with Meta-Analysis of Experimental and Observational Studies."
19. S. Amagasa et al., "Is Objectively Measured Light-Intensity Physical Activity Associated with Health Outcomes after Adjustment for Moderate-to-Vigorous Physical Activity in Adults? A Systematic Review," *Int J Behav Nutr Phys Act* 15, no. 1 (2018): 65.
20. R. A. Mekary et al., "Isotemporal Substitution Paradigm for Physical Activity Epidemiology and Weight Change," *Am J Epidemiol* 170, no. 4 (2009): 519–27.
21. S. F. Chastin et al., "Combined Effects of Time Spent in Physical Activity, Sedentary Behaviors and Sleep on Obesity and Cardio-Metabolic Health Markers: A Novel Compositional Data Analysis Approach," *PLoS One* 10, no. 10 (2015): e0139984.
22. Brocklebank et al., "Accelerometer-Measured Sedentary Time and Cardiometabolic Biomarkers: A Systematic Review."
23. J. Grgic et al., "Health Outcomes Associated with Reallocations of Time between Sleep, Sedentary Behaviour, and Physical Activity: A Systematic Scoping Review of Isotemporal Substitution Studies," *Int J Behav Nutr Phys Act* 15, no. 1 (2018): 69.

24. A. García-Hermoso et al., "Reallocating Sedentary Time to Moderate-to-Vigorous Physical Activity but Not to Light-Intensity Physical Activity Is Effective to Reduce Adiposity among Youths: A Systematic Review and Meta-Analysis," *Obes Rev* 18, no. 9 (2017): 1088–95.
25. I. Janssen et al., "A Systematic Review of Compositional Data Analysis Studies Examining Associations between Sleep, Sedentary Behaviour, and Physical Activity with Health Outcomes in Adults," *Appl Physiol Nutr Metab* 45, no. 10 (2020): S248–S57.
26. B. B. Gibbs et al., "Cross-Sectional and Longitudinal Associations between Objectively Measured Sedentary Time and Metabolic Disease: The Coronary Artery Risk Development in Young Adults (CARDIA) Study," *Diabetes Care* 38, no. 10 (2015): 1835–43.
27. C. A. Celis-Morales et al., "Objective vs. Self-Reported Physical Activity and Sedentary Time: Effects of Measurement Method on Relationships with Risk Biomarkers," *PLoS One* 7, no. 5 (2012): e36345.
28. M. P. Buman et al., "Reallocating Time to Sleep, Sedentary Behaviors, or Active Behaviors: Associations with Cardiovascular Disease Risk Biomarkers, NHANES 2005–2006," *Am J Epidemiol* 179, no. 3 (2014): 323–34.
29. K. H. Herzig et al., "Light Physical Activity Determined by a Motion Sensor Decreases Insulin Resistance, Improves Lipid Homeostasis and Reduces Visceral Fat in High-Risk Subjects: Prediabex Study RCT," *Int J Obes* 38, no. 8 (2014): 1089–96.
30. M. Pyörälä et al., "Plasma Insulin and All-Cause, Cardiovascular, and Noncardiovascular Mortality: The 22-Year Follow-up Results of the Helsinki Policemen Study," *Diabetes Care* 23, no. 8 (2000): 1097--102.
31. V. Carson et al., "Patterns of Sedentary Time and Cardiometabolic Risk among Canadian Adults," *Prev Med* 65 (2014): 23–27.
32. G. N. Healy et al., "Sedentary Time and Cardio-Metabolic Biomarkers in US Adults: Nhanes 2003–06," *Eur Heart J* 32, no. 5 (2011): 590–97.
33. B. M. Lynch et al., "Associations of Objectively Assessed Physical Activity and Sedentary Time with Biomarkers of Breast Cancer Risk in Postmenopausal Women: Findings from Nhanes (2003–2006)," *Breast Cancer Res Treat* 130, no. 1 (2011): 183–94.
34. C. J. Brakenridge et al., "Associations of 24 H Time-Use Compositions of Sitting, Standing, Physical Activity and Sleeping with Optimal Cardiometabolic Risk and Glycaemic Control: The Maastricht Study," *Diabetologia* 67, no. 7 (2024): 1356–67.
35. U. Ekelund et al., "Does Physical Activity Attenuate, or Even Eliminate, the Detrimental Association of Sitting Time with Mortality? A Harmonised Meta-Analysis of Data from More Than 1 Million Men and Women," *Lancet* 388, no. 10051 (2016): 1302–10.
36. B. M. F. M. Duvivier et al., "Minimal Intensity Physical Activity (Standing and Walking) of Longer Duration Improves Insulin Action and Plasma Lipids More Than Shorter Periods of Moderate to Vigorous Exercise (Cycling) in Sedentary Subjects When Energy Expenditure Is Comparable," *PLoS One* 8, no. 2 (2013): e55542.
37. U. Ekelund et al., "Dose-Response Associations between Accelerometry Measured Physical Activity and Sedentary Time and All Cause Mortality: Systematic Review and Harmonised Meta-Analysis," *BMJ* 366 (2019): l4570.
38. F. R. Cavallo et al., "The Association between Sedentary Behaviour, Physical Activity and Type 2 Diabetes Markers: A Systematic Review of Mixed Analytic Approaches," *PLoS One* 17, no. 5 (2022): e0268289.
39. Cavallo et al., "The Association between Sedentary Behaviour, Physical Activity and Type 2 Diabetes Markers: A Systematic Review of Mixed Analytic Approaches."
40. C. Powell et al., "The Cross-Sectional Associations between Objectively Measured Sedentary Time and Cardiometabolic Health Markers in Adults – A Systematic Review with Meta-Analysis Component," *Obes Rev* 19, no. 3 (2018).

41. V. J. Silfee et al., "Objective Measurement of Physical Activity Outcomes in Lifestyle Interventions among Adults: A Systematic Review," *Prev Med Rep* 11 (2018): 74–80.
42. C. L. Craig et al., "International Physical Activity Questionnaire: 12-Country Reliability and Validity," *Med Sci Sports Exerc* 35, no. 8 (2003):1381–95.
43. K. P. Dowd et al., "A Systematic Literature Review of Reviews on Techniques for Physical Activity Measurement in Adults: A Dedipac Study," *Int J Behav Nutr Phys Act* 15, no. 1 (2018): 15.
44. M. Fogelholm et al., "International Physical Activity Questionnaire: Validity against Fitness," *Med Sci Sports Exerc* 38, no. 4 (2006): 753.
45. K. Meh et al., "The Dilemma of Physical Activity Questionnaires: Fitter People Are Less Prone to Over-reporting," *PLoS One* 18, no. 8 (2023): e0285357.
46. Brocklebank et al., "Accelerometer-Measured Sedentary Time and Cardiometabolic Biomarkers: A Systematic Review."
47. L. Grimes et al., "Accelerometery as a Measure of Modifiable Physical Activity in High-Risk Elderly Preoperative Patients: A Prospective Observational Pilot Study," *BMJ Open* 9, no. 11 (2019): e032346.
48. Dowd et al., "A Systematic Literature Review of Reviews on Techniques for Physical Activity Measurement in Adults: A Dedipac Study."
49. R. M. Pulsford et al., "The Impact of Selected Methodological Factors on Data Collection Outcomes in Observational Studies of Device-Measured Physical Behaviour in Adults: A Systematic Review," *Int J Behav Nutr Phys Act* 20 (2023): 26.
50. Pulsford et al., "The Impact of Selected Methodological Factors on Data Collection Outcomes in Observational Studies of Device-Measured Physical Behaviour in Adults: A Systematic Review."
51. S. Verhoog et al., "Comparison of Different Software for Processing Physical Activity Measurements with Accelerometry," *Sci Rep* 13, no. 1 (2023): 2879.
52. S. A. Prince et al., "A Comparison of Self-Reported and Device Measured Sedentary Behaviour in Adults: A Systematic Review and Meta-Analysis," *Int J Behav Nutr Phys Act* 17, no. 1 (2020): 31.
53. S. Skender et al., "Accelerometry and Physical Activity Questionnaires – A Systematic Review," *BMC Public Health* 16, no. 1 (2016): 515.
54. Prince et al., "A Comparison of Self-Reported and Device Measured Sedentary Behaviour in Adults: A Systematic Review and Meta-Analysis."
55. P. W. Ku et al., "A Cut-Off of Daily Sedentary Time and All-Cause Mortality in Adults: A Meta-Regression Analysis Involving More Than 1 Million Participants," *BMC Med* 16, no. 1 (2018): 1–9.
56. R. Patterson et al., "Sedentary Behaviour and Risk of All-Cause, Cardiovascular and Cancer Mortality, and Incident Type 2 Diabetes: A Systematic Review and Dose Response Meta-Analysis," *Eur J Epidemiol* 33, no. 8 (2018): 811–29.
57. Dowd et al., "A Systematic Literature Review of Reviews on Techniques for Physical Activity Measurement in Adults: A Dedipac Study."
58. A. Elhakeem et al., "Physical Activity, Sedentary Time, and Cardiovascular Disease Biomarkers at Age 60 to 64 Years," *J Am Heart Assoc* 7, no. 16 (2018): e007459.
59. T. Honda et al., "Objectively Measured Sedentary Time and Diabetes Mellitus in a General Japanese Population: The Hisayama Study," *J Diabetes Investig* 10, no. 3 (2019): 809–16.
60. J. Kim et al., "Objectively Measured Light-Intensity Lifestyle Activity and Sedentary Time Are Independently Associated with Metabolic Syndrome: A Cross-Sectional Study of Japanese Adults," *Int J Behav Nutr Phys Act* 10 (2013): 1–7.
61. S. Knaeps et al., "Independent Associations between Sedentary Time, Moderate-to-Vigorous Physical Activity, Cardiorespiratory Fitness and Cardio-Metabolic Health: A Cross-Sectional Study," *PLoS One* 11, no. 7 (2016): e0160166.

62. C. M. Phillips, C. B. Dillon, and I. J. Perry, "Does Replacing Sedentary Behaviour with Light or Moderate to Vigorous Physical Activity Modulate Inflammatory Status in Adults?," *Int J Behav Nutr Phys Act* 14, no. 1 (2017): 138.
63. K. B Watson et al., "Comparison of Accelerometer Cut Points to Estimate Physical Activity in US Adults," *J Sports Sci* 32 (2014): 660–69.
64. K.e Ellis et al., "Hip and Wrist Accelerometer Algorithms for Free-Living Behaviour Classification," *Med Sci Sports Exerc* 48, no. 5 (2016):933–40.
65. M. Willetts et al., "Statistical Machine Learning of Sleep and Physical Activity Phenotypes from Sensor Data in 96,220 UK Biobank Participants," *Sci Rep* 8 (2018): 7961.
66. V. Farrahi et al., "Accnet24: A Deep Learning Framework for Classifying 24-Hour Activity Behaviours from Wrist-Worn Accelerometer Data under Free-Living Environments," *Int J Med Inform* 172 (2023): 105004.
67. V. Nunavath et al., "Deep Learning for Classifying Physical Activities from Accelerometer Data," *Sensors (Basel)* 21, no. 16 (2021): 5564.
68. V. Farrahi, P. J. Collings, and . Oussalah, "Deep Learning of Movement Behavior Profiles and Their Association with Markers of Cardiometabolic Health," *BMC Med Inform Decis Mak* 24, no. 1 (2024): 74.
69. Willetts et al., "Statistical Machine Learning of Sleep and Physical Activity Phenotypes from Sensor Data in 96,220 UK Biobank Participants."
70. V. Farrahi et al., "Calibration and Validation of Accelerometer-Based Activity Monitors: A Systematic Review of Machine-Learning Approaches," *Gait Posture* 68 (2019): 285–99.
71. D. Trabelsi et al., "An Unsupervised Approach for Automatic Activity Recognition Based on Hidden Markov Model Regression," *IEEE Trans. Autom. Sci. Eng.* 10 (2013): 829–35.
72. Willetts et al., "Statistical Machine Learning of Sleep and Physical Activity Phenotypes from Sensor Data in 96,220 UK Biobank Participants."
73. A. P. Creagh et al., "Digital Health Technologies and Machine Learning Augment Patient Reported Outcomes to Remotely Characterise Rheumatoid Arthritis," *NPJ Digit Med* 7, no. 1 (2024): 33.
74. H. Haresamudram, I. Essa, and T. Plötz, "Assessing the State of Self-Supervised Human Activity Recognition Using Wearables," *Proc ACM Interact Mob Wearable Ubiquitous Technol* 6, no. 3 (2022): 1–47.
75. W. H. Ong, T. Koseki, and L. Palafox, "An Unsupervised Approach for Human Activity Detection and Recognition," *Int J Syst Sci* 14 (2013): 42–49.
76. P. Jones et al., "Towards a Portable Model to Discriminate Activity Clusters from Accelerometer Data," *Sensors* 19, no. 20 (2019): 4504.
77. S. Clark et al., "Clustering Accelerometer Activity Patterns from the UK Biobank Cohort," *Sensors* 21, no. 24 (2021): 8220.
78. N. Weber, "Unsupervised Learning in Human Activity Recognition: A First Foray into Clustering Data Gathered from Wearable Sensors," (2017).
79. T. V. Duong et al., "Activity Recognition and Abnormality Detection with the Switching Hidden Semi-Markov Model," *Proc IEEE Comput Soc Conf Comput Vis Pattern Recognit (CVPR'05)* 1 (2005): 838–45.
80. Trabelsi et al., "An Unsupervised Approach for Automatic Activity Recognition Based on Hidden Markov Model Regression."
81. T. L. M. Van Kasteren, G. Englebienne, and B. J. A. Kröse, "Activity Recognition Using Semi-Markov Models on Real World Smart Home Datasets," *J Ambient Intell Smart Environ* 2 (2010): 311–25.
82. Van Kasteren, Englebienne, and Kröse, "Activity Recognition Using Semi-Markov Models on Real World Smart Home Datasets."

83. D. Van Kuppevelt et al., "Segmenting Accelerometer Data from Daily Life with Unsupervised Machine Learning," *PLoS One* 14 (2019): e0208692.
84. C. B. Thornton, N. Kolehmainen, and K. Nazarpour, "Using Unsupervised Machine Learning to Quantify Physical Activity from Accelerometry in a Diverse and Rapidly Changing Population," *PLOS Digit Health* 2, no. 4 (2023): e0000220.
85. R. Bender, "Introduction to the Use of Regression Models in Epidemiology," in *Cancer Epidemiology*, ed. Mukesh Verma, 179–95 (Totowa, NJ: Humana Press, 2009).
86. J. H. Migueles et al., "Granada Consensus on Analytical Approaches to Assess Associations with Accelerometer-Determined Physical Behaviours (Physical Activity, Sedentary Behaviour and Sleep) in Epidemiological Studies," *Br J Sports Med* 56 (2021): 376–84.
87. Ž. Pedišić, D. Dumuid, and T. S. Olds, "Integrating Sleep, Sedentary Behaviour, and Physical Activity Research in the Emerging Field of Time-Use Epidemiology: Definitions, Concepts, Statistical Methods, Theoretical Framework, and Future Directions," *Kinesiology* 49, no. 2. (2017): 252–69.
88. D. Dumuid et al., "Compositional Data Analysis for Physical Activity, Sedentary Time and Sleep Research," *Stat Methods Med Res* 27 (2018): 3726–38.
89. Chastin et al., "Combined Effects of Time Spent in Physical Activity, Sedentary Behaviors and Sleep on Obesity and Cardio-Metabolic Health Markers: A Novel Compositional Data Analysis Approach."
90. Mekary et al., "Isotemporal Substitution Paradigm for Physical Activity Epidemiology and Weight Change."
91. W. Willett, "Implications of Total Energy Intake for Epidemiologic Analyses," *Nutritional epidemiology* (1998): 273–301.
92. Dumuid et al., "Compositional Data Analysis for Physical Activity, Sedentary Time and Sleep Research."
93. G. J. H. Biddle et al., "Associations of Physical Behaviours and Behavioural Reallocations with Markers of Metabolic Health: A Compositional Data Analysis," *Int J Environ Res Public Health* 15, no. 10 (2018): 2280.
94. Brakenridge et al., "Associations of 24 H Time-Use Compositions of Sitting, Standing, Physical Activity and Sleeping with Optimal Cardiometabolic Risk and Glycaemic Control: The Maastricht Study."
95. D. Dumuid et al., "Adiposity and the Isotemporal Substitution of Physical Activity, Sedentary Time and Sleep among School-Aged Children: A Compositional Data Analysis Approach," *BMC Public Health* 18, no. 1 (2018): 311.
96. Pedišić, Dumuid, and Olds, "Integrating Sleep, Sedentary Behaviour, and Physical Activity Research in the Emerging Field of Time-Use Epidemiology: Definitions, Concepts, Statistical Methods, Theoretical Framework, and Future Directions."
97. Migueles et al., "Granada Consensus on Analytical Approaches to Assess Associations with Accelerometer-Determined Physical Behaviours (Physical Activity, Sedentary Behaviour and Sleep) in Epidemiological Studies."
98. Cavallo et al., "The Association between Sedentary Behaviour, Physical Activity and Type 2 Diabetes Markers: A Systematic Review of Mixed Analytic Approaches."
99. Migueles et al., "Granada Consensus on Analytical Approaches to Assess Associations with Accelerometer-Determined Physical Behaviours (Physical Activity, Sedentary Behaviour and Sleep) in Epidemiological Studies."
100. Dumuid et al., "Adiposity and the Isotemporal Substitution of Physical Activity, Sedentary Time and Sleep among School-Aged Children: A Compositional Data Analysis Approach."
101. Biddle et al., "Associations of Physical Behaviours and Behavioural Reallocations with Markers of Metabolic Health: A Compositional Data Analysis."

102. A. Miatke et al., "The Association between Reallocations of Time and Health Using Compositional Data Analysis: A Systematic Scoping Review with an Interactive Data Exploration Interface," *Int J Behav Nutr Phys Act* 20, no. 1 (2023): 127.
103. Cavallo et al., "The Association between Sedentary Behaviour, Physical Activity and Type 2 Diabetes Markers: A Systematic Review of Mixed Analytic Approaches."
104. Lynch et al., "Associations of Objectively Assessed Physical Activity and Sedentary Time with Biomarkers of Breast Cancer Risk in Postmenopausal Women: Findings from Nhanes (2003–2006)."
105. Healy et al., "Sedentary Time and Cardio-Metabolic Biomarkers in US Adults: Nhanes 2003–06."
106. Honda et al., "Objectively Measured Sedentary Time and Diabetes Mellitus in a General Japanese Population: The Hisayama Study."
107. Ekelund et al., "Dose-Response Associations between Accelerometry Measured Physical Activity and Sedentary Time and All Cause Mortality: Systematic Review and Harmonised Meta-Analysis."
108. Ku et al., "A Cut-Off of Daily Sedentary Time and All-Cause Mortality in Adults: A Meta-Regression Analysis Involving More Than 1 Million Participants."
109. G. Leandro et al., "Non-Occupational Physical Activity and Risk of Cardiovascular Disease, Cancer and Mortality Outcomes: A Dose–Response Meta-Analysis of Large Prospective Studies," *Br J Sports Med* 57, no. 15 (2023): 979.
110. W. Geidl et al., "Dose–Response Relationship between Physical Activity and Mortality in Adults with Noncommunicable Diseases: A Systematic Review and Meta-Analysis of Prospective Observational Studies," *Int J Behav Nutr Phys Act* 17, no. 1 (2020): 109.
111. K. F. Arnold et al., "A Causal Inference Perspective on the Analysis of Compositional Data," *International Journal of Epidemiology* 49, no. 4 (2020): 1307–13.
112. S. Chastin et al., "Joint Association between Accelerometry-Measured Daily Combination of Time Spent in Physical Activity, Sedentary Behaviour and Sleep and All-Cause Mortality: A Pooled Analysis of Six Prospective Cohorts Using Compositional Analysis," *Br J Sports Med* 55, no. 22 (2021): 1277–85; Arnold et al., "A Causal Inference Perspective on the Analysis of Compositional Data."
113. Wirth et al., "Biomarkers Associated with Sedentary Behaviour in Older Adults: A Systematic Review."
114. B. Silva et al., "Associations of Physical Activity and Sedentary Time with Body Composition in Brazilian Young Adults," *Sci Rep* 9, no. 1 (2019): 5444.
115. F. R. Cavallo, C. Toumazou, and K.n Nikolic, "Unsupervised Classification of Human Activity with Hidden Semi-Markov Models," *Appl Syst Innov* 5, no. 4 (2022): 83.
116. A. Doherty et al., "Large Scale Population Assessment of Physical Activity Using Wrist Worn Accelerometers: The UK Biobank Study," *PLoS One* 12, no. 2 (2017): e0169649.
117. Willetts et al., "Statistical Machine Learning of Sleep and Physical Activity Phenotypes from Sensor Data in 96,220 UK Biobank Participants."
118. M. J. Johnson and A. S. Willsky, "Bayesian Nonparametric Hidden Semi-Markov Models," *J Mach Learn Res* 14, no. 1 (2013): 673–701.
119. Y. W.e Teh, "Dirichlet Process," *Encycl Mach Learn* 1063 (2010), 280–287.https://citeseerx.ist.psu.edu/document?repid=rep1&type=pdf&doi=71cc17b9f3ffeaaf7dc69db141cc74c1f39bc8dd
120. V. Witowski et al., "Using Hidden Markov Models to Improve Quantifying Physical Activity in Accelerometer Data – A Simulation Study," *PLoS One* 9 (2014): e114089.
121. Johnson and Willsky, "Bayesian Nonparametric Hidden Semi-Markov Models."
122. Willetts et al., "Statistical Machine Learning of Sleep and Physical Activity Phenotypes from Sensor Data in 96,220 UK Biobank Participants."

123. Willetts et al., "Statistical Machine Learning of Sleep and Physical Activity Phenotypes from Sensor Data in 96,220 UK Biobank Participants."
124. World Health Organization, "Global Health Risks: Mortality and Burden of Disease Attributable to Selected Major Risks," World Health Organization, Geneva (2009), https://www.who.int/publications/i/item/9789241563871
125. P. Slade et al., "Sensing Leg Movement Enhances Wearable Monitoring of Energy Expenditure," *Nat Commun* 12, no. 1 (2021): 4312.
126. P. Slade et al., "An Open-Source and Wearable System for Measuring 3D Human Motion in Real-Time," *IEEE Trans Biomed Eng* 69, no. 2 (2021): 678–88.
127. Cavallo et al., "The Association between Sedentary Behaviour, Physical Activity and Type 2 Diabetes Markers: A Systematic Review of Mixed Analytic Approaches."
128. C. Sudlow et al., "UK Biobank: An Open Access Resource for Identifying the Causes of a Wide Range of Complex Diseases of Middle and Old Age," *PLoS Medicine* 12 (2015): 1–10.
129. Doherty et al., "Large Scale Population Assessment of Physical Activity Using Wrist Worn Accelerometers: The UK Biobank Study."
130. Willetts et al., "Statistical Machine Learning of Sleep and Physical Activity Phenotypes from Sensor Data in 96,220 UK Biobank Participants."
131. Chastin et al., "Combined Effects of Time Spent in Physical Activity, Sedentary Behaviors and Sleep on Obesity and Cardio-Metabolic Health Markers: A Novel Compositional Data Analysis Approach."
132. A. Doherty et al., "Gwas Identifies 14 Loci for Device-Measured Physical Activity and Sleep Duration," *Nat Commun* 9 (2018): 5257.
133. R. de Ritter et al., "Adverse Differences in Cardiometabolic Risk Factor Levels between Individuals with Pre-Diabetes and Normal Glucose Metabolism Are More Pronounced in Women Than in Men: The Maastricht Study," *BMJ Open Diabetes Res Care* 7, no. 1 (2019): e000787.
134. J. Cohen, *Statistical Power Analysis for the Behavioral Sciences*, 2nd ed. (New York: Routledge, 1988).
135. Y. Rosseel, "Lavaan: An R Package for Structural Equation Modeling," *J Stat Softw* 48, no. 2 (2012): 1–36.
136. Doherty et al., "Gwas Identifies 14 Loci for Device-Measured Physical Activity and Sleep Duration."

3 Digital technologies for personalised behavioural change

PERSONALISED BEHAVIOURAL CHANGE

Tackling global health issues depends on changing human behaviour. There is overwhelming evidence showing that health behaviours can significantly influence diseases contributing to increased mortality and morbidity: the highest proportion of all-cause disability-adjusted years is due to behavioural risks,[1] such as low physical activity, high levels of sedentary behaviour, poor diet, and smoking. Therefore, changing human behaviour can significantly alter the currently increasing trends of lifestyle-triggered diseases. At first glance, behaviour may seem the easiest factor to influence, given that other factors such as genetic predisposition, social circumstances, and environmental factors are difficult, if not impossible, to change. However, changing behaviour is a challenging task. For example, despite the well-known benefits of exercise, many people fail to follow national guidelines around physical activity. In the US, only 47% of adults meet the physical activity guidelines for aerobic physical activity, and only 24% meet the guidelines for aerobic and muscle-strengthening activities.[2] The UK fares slightly better, with 63% of adults meeting the guidelines of at least 150 minutes of activity per week.[3]

Many attempts at changing health-related behaviours, such as reducing alcohol consumption or increasing physical activity, have proven largely unsuccessful.[4] There are many reasons behind this, but perhaps the most significant is the over-simplification of human behaviour and its drivers. Mistakes commonly found behind unsuccessful public health campaigns include assuming that knowledge and information are the only drivers of behaviour, or that people act fully rationally—or fully irrationally—or that it is possible to predict patterns of behaviour accurately.[5] Health behaviours are determined by a multitude of factors, including biological, psychosocial, cultural, environmental, and institutional, and therefore are understandably challenging to change. Deeply understanding which factors influence behaviour, and how, is the key to behaviour change. Several theoretical models try to explain what drives health behaviours, from the Health Belief Model[6] describing behaviour as being driven by the combination of the risk associated with the behaviour and the effort required to change it, to the Information, Motivation, and Behavioural Skills Model[7] stating that individuals must be informed, motivated and behaviourally skilled to engage in behavioural change. These theories highlight the

DOI: 10.1201/9781003464983-3

importance of knowledge to drive change, for example, around the risks of certain behaviours. The Theory of Planned Behaviour[8] describes behaviour more holistically as being driven by attitude, intentions, subjective and social norms, perceived power, and perceived behavioural control. One of the most recent and perhaps systematic models of behaviour is the Capability–Opportunity–Motivation Model of Behaviour,[9] according to which a person must be physically and psychologically able to exhibit the behaviour, have the opportunity to do so, and possess the desire or need to demonstrate the behaviour at that moment. Health behaviour theories are covered in more detail in the next section, titled "Behavioural theory to understand and change behaviour."

Given the multitude of factors that affect human behaviour, personalised interventions hold considerable promise. The use of personalisation paired with behavioural theory has been positively correlated with successfully changing behaviours around physical activity[10] and diet,[11] among others. Interventions that are closely aligned to the individual's motivations are more likely to be observed and remembered and, therefore, personalisation can overcome the lack of adherence and increase effectiveness.[12] Personalisation can improve a person's perceived skills and their ability to succeed, also known as self-efficacy, and motivation.[13] In fact, according to the Self-Efficacy Theory,[14] motivation is driven by self-efficacy, which is developed through repeated successes to challenges that become more difficult as the individual's skills improve. Personalising the intervention is an effective way to progressively build high self-efficacy, which is key in successfully enacting the behaviour change process and maintaining the new behaviour in the long term.

There are several ways in which personalisation can be implemented, and different types of personal data can be used to realise personalised interventions. There has been extensive research into the effectiveness of genetic risk communication with mixed results. Some have found it to be effective in encouraging dietary changes,[15,16] while other studies and reviews could not find benefits of communicating the genetic predisposition to diseases to change health behaviours.[17,18] Several reviews stated that lacking and poor-quality evidence precludes drawing definitive conclusions.[19,20] Lack of evidence is not the only factor against genotype-based personalised advice to promote behaviour change: studies show that genetic risk communication could also increase anxiety and demotivation.[21] However, another review[22] found that studies that compared genetic risk communication to general health advice found improvements, but those improvements are not observed when genetic personalisation is compared to other forms of personalisation. This highlights that while genetic information does not bring further benefits, the concept of personalisation is effective when compared to non-tailored advice. But again, the body of evidence is not coherent, and a randomised controlled trial (RCT) assessing the impact of communicating personalised disease risk to increase physical activity had no success.[23]

As well as genetic data, biomarkers have been used to communicate disease risk with the hope of encouraging individuals to implement behaviour change. However, this has been largely unsuccessful, as shown by a systematic review that did not find sufficient evidence to support the hypothesis that biomarker-based risk communication increases smoking cessation when compared to standard treatments.[24]

Several behavioural models identify risk perception and knowledge as key to behaviour, and, at least theoretically, personalised risk communication should improve behaviour change, as it leads to a better understanding of health risk[25] and to a greater acceptance of the message regarding risk.[26] So why has this strategy been so ineffective at driving change? A review of systematic reviews[27] found little evidence that personalised risk information has a strong effect on health-related behaviours. It was also found that there was often no behavioural theory underpinning the interventions, despite widespread knowledge that targeting self-efficacy increases the impact of risk information on behaviour. The authors also found that many of the studies included in the reviews were of poor quality. In terms of efficacy, they found that different data sources for personalisation achieved different results, with imaging-based approaches being the most effective. Numerical-based approaches were found to be the least helpful, which may explain why risk scores based on genetics have generally been unhelpful. However, the authors stress that personalised risk communication may only be effective in some specific cases which, paired with the lack of theoretical rationales and quality, may explain the lack of consistent results in the literature.[28] Notably, when studies incorporated behavioural theory in their design, they reported positive findings.[29,30,31] A more recent review[32] published in 2018 replicated the same findings as previous reviews: studies evaluating the efficacy of genetic risk communication are generally of poor quality, and they do not use behaviour change theory to inform the study design or analysis, thus being prone to several confounding factors such as attitudes, subjective norms, and self-efficacy. A systematic review published in 2023[33] conducted a meta-analysis that showed genotype-based advice did not affect dietary or physical activity behaviour more than general advice or advice based on lifestyle or phenotypic aimed at behaviour change to reduce the risk of cardiovascular disease (CVD), type 2 diabetes (T2D), and obesity. With these considerations, it is perhaps unsurprising that many studies fail to prove the benefits of personalised risk communication.

Risk communication alone seems largely ineffective, and research has shown that simply giving people information is not enough to change behaviour.[34] When personal information is used more holistically to tailor actionable recommendations, behaviour change can be positively impacted. For example, an RCT[35] found that actionable recommendations tailored with nutritional genetic information led to reduced fat intake both in the short and long term; a reduction that was not seen in the control group. Another dietary intervention personalised on individual glucose responses was better at stabilising blood glucose levels compared to the control group.[36]

Digital technologies can enable more sophisticated forms of personalisation, which are also more effective in changing behaviours. For example, web-based personalised interventions are more effective than non-personalised advice in reducing body weight and increasing fruit and vegetable consumption[37] and in general, the internet is a feasible and acceptable platform for delivering effective and large-scale lifestyle-based interventions.[38] A systematic review and meta-regression[39] found that personalised interventions delivered through mobile technologies are effective in changing lifestyle behaviours around physical activity, diet, smoking, and alcohol consumption. Technology-enabled interventions can be especially useful in increasing activity and reducing sedentary time, thus improving glucose control in T2D patients.[40] Just-in-time adaptive interventions (JITAIs) hold great potential to enable behaviour change, as they can deliver the intervention remotely and "at the moment and in the context that the person needs it most and is most likely to be receptive."[41] Thanks to the widespread availability of mobile technology, JITAIs can be implemented at scale with the potential for significant impact on disease management and prevention. However, despite the promising premise and initial results,[42] research into JITAIs is in its infancy.[43,44]

Despite mixed evidence, personalisation holds great promise to aid successful behaviour change, and the extent to which personalisation is effective in behavioural interventions depends on the integration of behavioural change theory and technology. On the one hand, understanding the individual drivers of behaviour through behavioural theory can inform what techniques work best for a certain behaviour, population, and environment. On the other hand, technology can create interventions that are automated, engaging, and truly personalised.

BEHAVIOURAL THEORY TO UNDERSTAND AND CHANGE BEHAVIOUR

There is a general consensus that behavioural change interventions should be underpinned by evidence and robust theoretical frameworks to increase their effectiveness.[45,46] The integration of evidence-based strategies ensures that the interventions are grounded in proven methodologies, thereby increasing the likelihood of successful health outcomes. In particular, a meta-analysis showed that mobile-based physical activity interventions are more effective when they are designed using behavioural change theories and models.[47] These theories and models provide a structured approach to understanding and influencing health behaviours, which is crucial for designing interventions that can effectively motivate individuals to adopt and maintain healthier lifestyles. By leveraging these theoretical frameworks, developers and healthcare providers can create more targeted, personalised, and impactful interventions that sustain long-term behaviour change, ultimately contributing to improved health outcomes.

Several health behaviour theories have been used to inform health intervention designs and to explain health behaviours, such as the Health Belief Model,[48] the Theory of Planned Behaviour,[49] the Health Action Process approach,[50] the Information, Motivation, and Behavioural Skills Model,[51] the Transtheoretical Model,[52] and the Capability–Opportunity–Motivation Model.[53]

One of the very first models of behavioural change, the Health Belief Model (HBM), was first introduced in the 1950s to understand why people failed to participate in programmes for disease prevention and immunisation. It describes behaviour as driven by a combination of six factors: the perceived risk associated with the behaviour, the perception of the severity of the health condition, the perceived benefit of engaging in the positive behaviour, the perceived barrier or effort required to change the damaging behaviour, the cue to action, and self-efficacy, i.e., the confidence in one's ability to perform the behaviour change. If the perception of the risk is high and the individual thinks they will benefit from changing habits, they are more likely to initiate the behavioural change. The HBM has been successfully used to design health interventions, for example, to improve behavioural risks associated with CVD,[54] and to explain health-related behaviours such as weight management.[55] However, the HBM suffers from several limitations. Firstly, it does not account for a person's attitudes, beliefs, and motivation; secondly, it does not consider the effect of social, environmental, and economic factors that may prohibit or promote the behaviour change. As a result, the HBM has been found to only explain 20–40% of the variance in behaviours.[56] To address these limitations, over the years researchers have proposed various additions to the original HBM, with one of the latest extensions, proposed by Orji et al.,[57] adding four new variables as determinants of health behaviours: self-identity, perceived importance, consideration of future consequences, and concern for appearance. Their proposed extended model achieved a 78% increase in predictive capacity compared to the previous model when predicting healthy eating behaviour.

According to the Theory of Planned Behaviour (TPB), proposed by Ajzen in 1991, intention is the main factor determining behaviours. This intention is influenced by three factors: attitudes toward the behaviour, subjective norms such as perceived social pressure to perform or not perform the behaviour, and perceived behavioural control, i.e., the perceived ease or difficulty of performing the behaviour, which is similar to the concept of self-efficacy. These components interact to shape an individual's behavioural intentions and, therefore, their actions. The TPB has been widely applied in various fields, such as health psychology, to understand behaviours like smoking cessation,[58] physical activity,[59] and dietary choices.[60,61] Similar to the HBM, the TPB has been criticised for focusing exclusively on rational reasoning, whereby knowledge of the significance of changing risky behaviours is sufficient to engage in more healthy ones, and for excluding more unconscious influences on health behaviour,[62] including emotions.[63]

The Health Action Process Approach (HAPA) was proposed by Schwarzer in 1992[64] as an alternative to previous models, including the HBM and the TPB. The HAPA suggests that initiating, adopting, and maintaining a health behaviour depends on a motivational phase, when the intention is created, and a volitional phase, when the intention is translated into action. During the motivational phase, three factors contribute to intention creation—self-efficacy, outcome expectancy, and perception of the risk associated with the behaviour—which can also interact and mediate each other. The volition phase can be further divided into a planning phase, action phase, and maintenance phase, emphasising the role of planning to bridge the intention–behaviour gap. Outcome expectancy is thought to be a precursor of self-efficacy, as individuals first review the pros and cons of engaging in the risky behaviour before contemplating whether they will be able to effectively engage in healthier behaviours. Self-efficacy is central in the HAPA, and it evolves along the motivation–volition–action chain. Firstly, the action self-efficacy, i.e., the perception that we will manage to achieve the change, contributes to the creation of the intention. Subsequently, the maintenance self-efficacy determines the ability to maintain the change despite adverse factors, both internal, such as temptation, and external, such as environmental barriers. Finally, the recovery self-efficacy affects the ability to recover after failures and, therefore, affects the likelihood of maintaining the behaviour change long-term, despite setbacks. This model has been used to design health behaviour change interventions, for example aimed at promoting physical activity and vegetable intake,[65] and to explain health behaviours such as motivation to quit smoking.[66] Despite addressing some of the issues associated with previously proposed behavioural models by including different types of motivation, the HAPA model of behaviour is missing important factors such as environment and social norms.

The Information, Motivation, and Behavioural Skills (IMB) model is another model proposed in the early 1990s by Fisher and Fisher,[67] commonly used to explain health behaviours. Initially developed to aid in the prevention of HIV/AIDS, it speculates that information, motivation, and behavioural skills are the fundamental determinants of health behaviour. Individuals have to be informed, motivated, and behaviourally skilled to engage in behavioural change. They need to know the risks of unhealthy behaviours and the benefits of healthier habits; they need to be motivated in multiple ways (for example, through goal setting, social engagement, or rewards); and they need to have the necessary skills. Similar to the HAPA, the IMB model emphasises that, while information is key to the behavioural change process, it is insufficient to drive change, and motivation plays an important part. The IMB model and various modified versions have been used to explain health-related behaviours such as medication adherence[68,69] and sexual health-related behaviours,[70,71] among others.

The Transtheoretical Model (TTM) was developed in the late 1970s to explain behaviour change in the context of smoking and is focused on the intention to change and the individual's decision-making process.[72] According to the TTM, an individual moves through five stages of change. During the precontemplation stage, the individual does not intend to take action in the foreseeable future and is not fully aware of the dangers that the behaviour in question involves. During the contemplation phase, the individual is aware of the negative consequences of the behaviour and is contemplating making a change in the foreseeable future. During the preparation phase, the individual is preparing to engage in the behaviour change in the near future and may start implementing the first steps towards change. The preparation phase is followed by the action phase, in which the individual actively engages in the new behaviour. During the maintenance phase, the individual has maintained the behaviour change for over six months and plans to continue sustaining it while focusing on strategies to avoid relapse to previous stages. The TTM has been extensively used to plan exercise-promoting interventions, but with limited success.[73] Reasons for this may be because exercise is a complex of different behaviours, while the TTM was developed for the well-defined one-activity behaviour that is smoking. Additionally, the TTM lacks the various external factors that influence complex behaviours like exercise.

The Capability–Opportunity–Motivation Behaviour (COM-B) model was developed by Michie et al.[74] and defines behavioural change as a combination of motivation, opportunity, and capability. Unlike in previously described models, motivation is considered both a conscious and an unconscious process—habits and emotions influence behaviour as much as goal setting and intention. However, motivation is not sufficient. Capability, both physical and psychological, and physical and social opportunity play an equally important role in determining whether behaviour change will be successful. Additionally, the COM-B model avoids particular emphasis on a single element of the behavioural change process and highlights the equal importance of external and internal factors. Together with the COM-B model, Michie et al. developed the Behavioural Change Wheel (BCW), a comprehensive, systematic, and evidence-based tool designed to aid the design of behaviour change interventions. The COM-B model and the associated BCW have been used to develop digital interventions that are both accepted by users and effective for a variety of behaviours, including improving diet,[75,76,77] increasing physical activity,[78,79,80,81] reducing sedentary behaviour,[82] managing hypertension[83] and diabetes,[84,85] smoking cessation,[86,87] alcohol reduction,[88] and medication adherence[89,90] or discontinuation.[91]

DIGITAL TECHNOLOGIES FOR BEHAVIOURAL CHANGE

Behavioural theories provide a foundational understanding of the factors that influence health behaviour change. In recent years, digital technologies have significantly impacted the way these theoretical principles are applied in real-world settings, improving the efficacy of behaviour change interventions. In particular, the new field of digital health has emerged, which refers to the use of information and communication technologies—such as mobile phones, apps and websites, remote monitoring, and smart wearables—to deliver healthcare, be they health interventions, illness management, or risk assessment.

In a report by the Institute of Global Health Innovation at Imperial College London,[92] the authors highlighted the potential for digital technologies to achieve universal health coverage, thanks to their ability to reach underserved populations, reduce costs, and improve the quality and safety of services. In addition, digital health enables a patient-centric approach, in which the individual's preferences, needs, and values are central to the provision of healthcare in order to achieve the best experience and outcome.[93] Digital technologies also move health delivery closer to the point of care and the patients, allowing individuals to feel more in control of their health, for example by retaining ownership over their data. At the same time, the richness of the data that can be collected through digital health provides a more holistic view of the patient's health, thus improving the quality of care.

Digital health tools are particularly suited to behavioural change. The inherently scalable nature of digital technologies means that they can reach large audiences in a more resource-effective way, making them particularly suitable for public health interventions. Additionally, these technologies can capture real-time data on individual behaviours and use it to deliver personalised feedback and intervention adjustment. They can facilitate the self-monitoring and feedback mechanisms highlighted in behavioural theories, helping individuals track their progress and adjust their behaviours accordingly. They can provide tailored interventions and offer targeted support. The integration of behavioural theory with digital technology not only improves the accessibility and scalability of health interventions but also enhances their efficacy and impact.[94] The usefulness of digital interventions is recognised by the recently published National Institute for Health and Care Excellence (NICE) guidelines on behaviour change,[95] which recommend the use of digital and mobile health tools to aid the change of unhealthy behaviours such as smoking, high alcohol consumption, unhealthy diet, sedentary behaviour, and unsafe sex.

Digital health encompasses many different technologies and use cases, and the Organisation for Economic Co-operation and Development (OECD) defines digital health as a combination of solutions including electronic health and medical records (EHR and EMR), telehealth, mobile health (mHealth), data-driven approaches such as artificial intelligence (AI) and machine learning (ML), and assisted-living technologies such as sensors and robots.[96]

EHRs and EMRs contain, at least in principle, the totality of an individual's health data in digital form. EMRs are used by a single medical organisation, while EHRs contain the individual's complete medical history, independent of which healthcare organisation the patient has interacted with. Despite their main aim being the improvement of quality and efficiency of care, EHRs and EMRs have often resulted in increased administrative burden for healthcare professionals[97] and decreased accuracy that can pose a real harm to patient safety.[98]

Telehealth and telemedicine refer to the use of communication technologies to enable remote medical care, consultations, monitoring, education, and support.[99] While home-based healthcare delivery has a long history,[100] the Covid-19 pandemic significantly accelerated the adoption of telephone and video consultations, providing an opportunity to adopt telehealth widely and consistently in order to reduce the burden on health systems and improve patient care.[101] Mobile health, or mHealth, refers to the use of mobile wireless technologies for public health. mHealth has been shown to increase access to health information, services, and skills, and to promote positive changes in health behaviours that prevent and manage diseases.[102] mHealth holds benefits both for the patients and healthcare providers. It enables patients to track health measurements, set medication and appointment reminders, access interventions and educational material, and consult their medical history, among other applications. For providers, mHealth offers the possibility to remotely track and communicate with patients and share health-related information using mobile devices. mHealth has the potential to revolutionise healthcare thanks to its widespread appeal, accessibility, and ability to reach large populations at a low-cost.[103]

Sensors and wearable devices are often used in conjunction with other technologies, including telehealth, to enable remote monitoring and care delivery. Remote monitoring also represents a significant tool to oversee the growing elderly population, who may face challenges in mobility and accessing healthcare services. For example, activity trackers have been used to monitor ambulatory disability in individuals with multiple sclerosis,[104] and remote support services for people with intellectual and developmental disabilities are increasingly being used to promote independence and reduce dependency on services.[105,106]

In recent years, AI and ML have attracted significant attention for their potential to transform healthcare. AI and ML can greatly enhance the design and delivery of personalised interventions through the use of predictive analytics, adaptive learning algorithms, and virtual assistants. AI can be used to tailor interventions based on individual needs and preferences, while predictive analytics can help identify at-risk individuals and predict future behaviour patterns, for example, medication adherence.[107] Chatbots and virtual assistants can provide real-time support, education, and encouragement for behaviour change.[108] Lately, virtual and augmented reality have been explored to create engaging and effective interventions to treat psychiatric conditions,[109] and to deliver physical therapy,[110] and psychological therapy.[111]

Despite the clear advantages that digital technologies present for healthcare and behavioural change, they also have challenges and limitations. The first significant concern is around data privacy and security. The opportunity that big data analytics offers also poses a threat to the security and privacy of patient information, while inadequate consent mechanisms open the possibility of patient data being sold to third parties. Over the years, there have been several notable examples of inadequate protection of identifiable patient information. For example, in 2017, the University of Chicago Medical Center and Google, which partnered to improve predictive analysis in medicine, were sued for failing to strip the data of information that could be used to identify patients.[112] In 2022, two healthcare systems in the US were sued over a data breach due to the use of third-party tracking software scraping patient information.[113] In 2021, an analysis of over 3,700 acute care hospitals in the US found that almost 99% of their websites track and transfer data to third parties, posing a significant risk to confidential patient data.[114] Cyberattacks and data breaches are not only a threat to the trust that patients have towards digital health, but can also have catastrophic implications for patients' health. The cyberattack that affected several UK hospitals in mid-2024 resulted in the cancellation of thousands of non-surgical appointments, hundreds of appointments for patients referred for urgent cancer diagnosis, cancelled C-sections, and transplant operations, as well as harming the ability to check patients' blood types.[115]

Another issue is around health equity. While the World Health Organization (WHO) identified digital health as a promising tool to reach universal health coverage,[116] the reliance on digital tools can exacerbate health disparities, as individuals without access to technology or with limited digital literacy may not benefit equally from digital health. To bridge the digital divide, digital health solutions need to be modelled on the needs of underserved communities, and their adoption should be accompanied by investments that address social determinants of health and increase access to the required technology.[117]

Several other issues arise depending on the nature of the technology. AI-enabled digital health opens the question of liability, which is especially evident in the case of misdiagnosis.[118] EHR management increases the administrative burden on physicians and reduces the time they can spend with patients.[119] The ease of distributing consumer-facing digital health solutions may mean that providers are able to escape regulation and avoid properly evaluating their products' quality, efficacy, and safety before launching to the market.[120,121,122]

Artificial intelligence

In past and current research, simple technologies including text messaging, web applications, and telehealth programmes have been popular tools for behavioural interventions to prevent or delay the onset of non-communicable diseases triggered by lifestyle practices, like T2D.[123] Recently, the growing popularity of new technologies to collect data, such as wearable trackers, has been offering new possibilities to increase the effectiveness of behaviour change interventions. For example, a meta-analysis found that enhanced interventions using computer, mobile, and wearable technologies can achieve a reduction in sedentary time of 42 minutes/day.[124] Specifically, AI and ML have emerged as transformative tools to help initiate and sustain behaviour change. By leveraging vast amounts of data obtained from various sources, including EHR, wearable devices, and questionnaires, AI and ML enable the creation of personalised and adaptive systems that guide and support the individual through the behaviour change process.

Despite mHealth being a promising tool to achieve behaviour change at scale, mHealth interventions have high abandonment rates, which have been associated with low perceived relevance and low engagement,[125] highlighting the need to explore additional motivators to encourage sustainable behaviour change. The persuasive power of behaviour change can be increased by personalising the intervention to meet the characteristics and preferences of each individual.[126] More advanced technologies can improve the effectiveness of behavioural interventions by creating adaptive and context-aware applications that are engaging and truly personalised. However, even newer technologies fail to achieve long-term behavioural change if only used for tracking behaviours. If interventions are not engaging, the users' adherence will drop once the novelty of the intervention wears off.[127] A direct solution to the issue of adherence is the creation of engaging interventions that respond to the environment and adapt to the user's needs, skills, and goals. These types of interventions have been defined as JITAIs, context-aware interventions, ecological momentary interventions (EMIs), or real-time interventions, which all describe systems able to deliver behaviour change support when the individual wants it or needs it the most.[128] Naughton[129] defines behavioural JITAIs as having three key features: they support the individual when they are at risk of engaging in a negative behaviour or they have the opportunity to engage in the positive one; they provide content and timing that is tailored to the individual based on data collected by the system; they provide support that is triggered by the system and not the user. JITAIs have been applied to several domains of health behaviour change, including physical activity,[130] alcohol use disorder,[131] and smoking cessation.[132]

AI and ML are powerful tools for creating context-aware and adaptive interventions individually tailored to increase adherence to programmes. For example, reinforcement learning, an ML technique in which an agent learns to make decisions by exploring possible actions and receiving rewards accordingly,[133] has been used to improve targeting of interventions for individuals with mild depression,[134] to optimise epilepsy treatment,[135] to improve adherence to diabetes medication,[136] and to increase physical activity in diabetes and depression. Predictive models have been used to forecast the outcome of interventions. For instance, Aswani et al.[137] proposed a Bayesian model to predict the outcome of weight loss programmes, with the aim of optimising therapies based on patients' characteristics. Ganguly et al.[138] used a bidirectional long short-term memory neural network to predict outcomes from RCTs for behaviour change, showing that AI can be used by policymakers or researchers implementing or designing behavioural change interventions for a target population. Moreover, several mHealth interventions based on conversational agents have been developed, including for exercise coaching[139] and weight loss.[140]

Recommender systems

Recommender systems are ML-based tools that predict the relevance of an item for a given user.[141] Depending on the approach used, recommender systems can be divided into content-based, collaborative filtering, or knowledge-based.[142] Content-based filtering makes recommendations relying only on the historical interactions that an individual has had with the system in the past in order to make recommendations; user-based collaborative filtering considers the similarity between users to make recommendations; and hybrid recommendation filters use a mix of content-based and collaborative filtering. Knowledge-based systems provide recommendations that are stored in a database and are not influenced by recent user preferences. Additionally, trust-based recommendation systems aim to improve the quality of the recommendation by considering the relationships between users,[143] while context-aware systems incorporate contextual information, such as time, location, or mood, to make recommendations under certain circumstances.[144]

Within healthcare, recommender systems have been used for a variety of applications. Clinical decision support systems that assist physicians during the decision-making process have been developed to recommend appropriate drug treatment,[145] personalised tuberculosis treatment,[146] diabetes treatment,[147] or to recommend the best physician for each person,[148] to name a few. Knowledge-based recommender systems are popular for health education applications; for instance, Rahdari et al.[149] designed a recommender system to provide education to ovarian cancer patients and carers.

Many health recommender systems are content-based, as they use real-time collected data to create a user profile in order to make future recommendations that support behaviour change. For instance, Ni et al.[150] built neural networks to recommend workout routes based on user data and expectations. They used a long short-term memory Recurrent Neural Network (RNN) trained on user identity, sport type, and historical workout sequences to recommend workout routes based on personal criteria, such as workout length, heart rate, and expectations. The model also predicts whether a user's heart rate will exceed a user-set threshold if they continue at the current pace and suggests ways to hit the user's target. Mahyari et al.[151] developed an intervention with a RNN to issue personalised recommendations with a high likelihood of being performed based on past exercise history. Zhao et al.[152] developed and tested the efficacy of a tree-like personalised gamification intervention that recommends activity time, and location based on the user's general information, personality, and daily activity data from a smartphone or wearable tracker.

Adaptive recommendations that use ML methods to issue personalised adaptive goals and recommendations based on user data have also been proposed. For example, Rabbi et al.[153] used a multi-armed bandit to recommend changes to existing repeated behaviours based on the user's context and data.

Collaborative recommender systems have been mainly developed to tailor messages as part of a behaviour change intervention. Ahsan et al.[154] developed a recommender algorithm to tailor messages to encourage smoking cessation based on the similarity between users who indicated that they realised a motivational effect after receiving a given group of motivational SMS messages. Notably, they used the TPB to design the intervention.

Several examples of "non-standard" recommender systems, i.e., not fitting within one of the collaborative or content-based subtypes, have been presented in the literature. Khwaja et al.[155] proposed a system recommending activities that improve a person's subjective wellbeing. Chen et al.[156] developed an expert system that recommends food products with a suitable nutritional value depending on one's genetic risk of developing metabolic conditions.

After the advent of transformers, first introduced in 2017,[157] Large Language Models (LLMs) could represent the new wave of recommender systems powering behavioural change. LLMs are advanced AI systems capable of understanding and generating outputs based on vast amounts of data, which have been used to generate health messages,[158] to predict disease risk,[159] and recommend CVD prevention strategies.[160] As of June 2024, Fitbit and Google Research are working on a system powered by LLMs that can deliver personalised coaching, actionable messages, and guidance, depending on personal health and fitness goals.[161] Academic and commercial research in healthcare applications of LLMs is very active and still in its infancy, and several issues, including hallucinations (producing plausible-sounding but incorrect information), privacy concerns, and interpretability, will need to be addressed before LLMs can be safely adopted in clinical practice.

DESIGNING DIGITAL SYSTEMS AS BEHAVIOURAL INTERVENTIONS

The success of mHealth interventions is highly dependent on the quality of their design. Following a structured framework for the development of mHealth interventions is essential to ensure that they are evidence-based, engaging, and effective. Building mHealth interventions on recognised frameworks also enables comprehensive evaluation and systematic assessment of which interventions work for which populations, ultimately advancing the knowledge base on digital interventions for behaviour change.

While there are several frameworks to develop digital interventions for behavioural change, the first step when designing a digital intervention is usually to specify the target behaviour, the target outcome, and the target audience.[162] For example, the target behaviour could be physical activity, with the target outcome being time spent in moderate-intensity activity, or weight loss, or VO_2max, etc. The target audience also significantly impacts the intervention design, as different populations will not only have different needs, but also different degrees of digital and health literacy.

Once the aims and populations have been defined, a literature review of the available evidence on the topic should be conducted. The review should provide an overview of existing relevant interventions and results on their efficacy, as well as barriers and facilitators of the target behaviour for the target outcome. Understanding why the target population fails at changing the target behaviour, and which behavioural change techniques are effective, are crucial data to inform and guide the intervention design.

The intervention should be supported by behavioural theory, which means firstly deciding on a behavioural model/theory and then modelling the intervention to that theory. Behavioural theory helps to identify the variables that need to be targeted in order to increase intervention effectiveness. Additionally, the use of theory helps to replicate interventions and to systematically review which behavioural change techniques are most effective for a specific behaviour and a specific population. However, choosing a suitable behavioural theory is challenging, partially due to a lack of guidance. The UK Medical Research Council highlights the importance of basing interventions on behavioural theories but does not provide guidance on how to choose the most appropriate one and how to apply it.[163] Davies et al.[164] suggest considering the target behaviour, population, and the wider context and mapping the concepts identified through the literature review to the existing theory that best captures those concepts—which is still considerably vague guidance. The lack of a structured framework primed the development of the Theoretical Domains Framework[165] (TDF) to systematise the process of mapping determinants of behaviours to 14 theoretical domains: knowledge, skills, social/professional role and identity, beliefs about capabilities, beliefs about consequences, motivation and goals, memory attention and decision processes, environmental context and resources, social influences, emotion, action planning, optimism, reinforcement, and intervention. Subsequently, the BCW[166] was developed as a synthesis of 19 frameworks of behaviour change in order to improve intervention design. The BCW provides a framework to develop complex interventions that is comprehensive, coherent, and linked to a behavioural change model, the COM-B model.

To design interventions, the BCW outlines three steps:[167] firstly, specifying and understanding the target behaviour by identifying theoretical domains that explain the behaviour—the TDF can be used as a tool to aid this identification step; secondly, identifying intervention options to target the theoretical domains that explain the behaviour; thirdly, selecting appropriate behaviour change techniques and implementation options. To simplify and standardise the last step, Michie et al. created a hierarchically structured taxonomy consisting of 93 distinctive, non-overlapping behaviour change techniques clustered into 16 groupings,[168] which was recently updated to include 237 behaviour change techniques organised into 20 groups.[169]

The BCW has been used to design interventions to change a wider array of health-related behaviours and has been reported in more than 5,000 published studies.[170] The case study in this chapter offers a step-by-step example of how the BCW can be employed to design personalised adaptive interventions.

Aside from the BCW, several other frameworks exist to design behavioural change interventions. Researchers may need to use multiple design frameworks when developing specific types of digital interventions. For example, JITAIs require making several design choices specific to adaptive interventions that are not explicitly included in the BCW. Frameworks for the design of adaptive interventions consider four parameters that can be set to design interventions:[171,172] decision points, intervention options, tailoring variables, and decision rules. A decision point is the time at which an intervention decision is made, which can happen at a specified time, time interval, or randomly. Often, decision points are selected to be at the time the user transitions from a state of vulnerability to the negative behaviour to a state of opportunity to engage in the positive target behaviour. Intervention options are possible treatments that can be employed at a given decision point, such as information, advice, feedback, support, and others; intervention options overlap with the behavioural change techniques described by the BCW. Tailoring variables represent data on the individual that can be used to tailor the intervention options to each person. Finally, decision rules are logic statements that link intervention options and tailoring variables. When designing a JITAI, it is also important to choose appropriate proximal outcomes, i.e., short-term goals, to encourage long-term engagement and prevent intervention fatigue. Proximal outcomes can be behavioural (e.g., measured by the interaction of the patient with the intervention), cognitive (e.g., perceiving the intervention as useful), or affective (e.g., trusting the intervention).[173] Choosing appropriate proximal outcomes is essential to continuously evaluate the intervention and ensure that its efficacy is sustained in the long term.

The person-based approach proposed by Yardley et al.[174] complements theory-based frameworks, such as the BCW. The aim of the person-based approach is to conduct iterative qualitative research during the development process to fully understand the perspective of the target users. This goes hand in hand with user engagement, which is advocated as a key aspect of developing digital interventions.[175,176] A person-based approach starts from the planning phase, achieved through consultation with experts, user groups, and other relevant stakeholders, and its development progresses with iterative feedback cycles to refine the intervention and meet user requirements. Other user-centred design frameworks[177,178] have been proposed to ensure that the intervention design takes into account the needs of the end user.

The choice of intervention options warrants further mention. Selecting appropriate behavioural change techniques is challenging, and due to the absence of standardisation and inadequate reporting, the literature is a collection of numerous studies from which robust conclusions cannot be drawn regarding what techniques are effective for particular behaviours. A comprehensive literature review done in the planning stage, together with a thorough understanding of the target behaviour, theoretical domains mapped to the behaviour, as well as user engagement, should identify the behavioural change techniques that are likely to be effective for the given use case.

Practically, there are several ways in which personalised mHealth interventions can aid the behaviour change process. For lifestyle-related behaviours such as healthy eating, physical activity, and sedentary behaviour, intervention using self-monitoring paired with at least one other behaviour change technique is more effective in changing habits compared to intervention not using self-monitoring.[179,180,181] To increase self-efficacy, self-monitoring can be paired with customisable goals, action planning, and feedback, which can include comparisons to past trends, other people's performance, or personal goals.[182] In addition to automatically recording behaviours without the need for user input, personalised adaptive interventions employing wearables or apps can track the changes in motivation that naturally occur during the behaviour change process due to external and internal circumstances.[183] The collected information can subsequently be used to modulate the difficulty of future challenges, thus increasing self-efficacy in a sustainable manner. Furthermore, as the environment affects motivation and can often be a barrier to maintenance, prompts or triggers personalised both in regard to content and timing can be used to encourage engagement with the behaviour.

Gamification, while not a behavioural change technique itself, could be effective in promoting and sustaining healthy behaviours,[184] especially if behavioural change techniques are incorporated within a gamification-based intervention. Existing behaviour change apps include multiple techniques, with feedback, monitoring, behaviour comparison, reward, and threat being the most common.[185] In particular, using gamification with goal settings and rewards has been found effective in increasing physical activity in patients from low-income neighbourhoods with a high risk of adverse cardiovascular events,[186] while gamification with team competition was found to increase physical activity among overweight and obese adults.[187]

Evaluating digital health interventions

Despite a plethora of available health applications—according to a report by IQVIA, there were over 350,000 health-related applications available in the major app stores[188]—most of them lack robust evaluation. Edwards et al.[189] evaluated 1,680 health and wellness apps and found that none were rigorously evaluated for potential benefits or harm. As a consequence, organisations continue to highlight the need for studies that follow patients for longer times, and that assess benefits, usability, retention, and cost-effectiveness.[190]

Systematic reviews have found that mHealth interventions for behaviour change generally lack thorough evaluation, with most studies relying on heuristic testing, usability testing, user evaluation, and pilot testing, rather than RCTs to evaluate the efficacy of behaviour change techniques.[191] A review on the effectiveness of mHealth interventions for obesity and diabetes treatment and self-management found that most studies employed a small sample size and did not use rigorous data collection and analytic approaches; moreover, most studies were pilots and presented limitations in their study design.[192] When considering mHealth interventions aimed at changing lifestyle behaviour, where the long-term sustainability of the behaviour is key to ensuring positive health outcomes, there is an insufficient number of studies evaluating the long-term effectiveness (>6 months) of interventions with no follow-up measurements,[193,194,195] which can be related to lack of long-term engagement with the intervention. Additionally, studies tend to be conducted in high-income countries,[196,197] thereby precluding the assessment of efficacy in populations with lower socioeconomic status or residing in developing countries.

Another area of rapid development is that of AI-based chatbot designs and interventions for promoting healthy behaviours. Despite a rapid increase in publications on the topic, standard measures for evaluating chatbot interventions and theory-guided chatbots are still lacking.[198,199]

Such paucity of rigorous studies can be attributed to a variety of reasons. Evaluating behavioural change interventions is technically challenging. Human behaviour is affected by a multitude of elements that are highly individual and time-dependent, such as motivation, habits, resources, and opportunities, resulting in a different likelihood of enacting a behavioural response at any given time.[200] The concept that the response to behaviour change techniques is context-dependent, as well as being dependent on individual characteristics such as age, gender, and socioeconomic status, can affect how individuals respond to interventions,[201] making evaluation particularly challenging. Additionally, reflecting the complexity of behaviour, mHealth interventions can themselves be highly complex, with several design factors to be optimised and assessed for usability, acceptability, and efficacy.

Evaluation requirements need to be considered during the developmental stages in order to allow the evaluation process,[202] which is at odds with the "ship fast, fail fast" mantra of the tech world. The rapidity at which the technological landscape changes makes conducting RCTs unrealistic, as they take an average of 5.5 years.[203] Additionally, several sources have argued that RCTs are likely not the most effective and pragmatic way to provide evidence of the benefits of digital health solutions.[204,205] even though they remain valuable for late-stage final assessments.[206]

Despite RCTs being considered the gold standard to evaluate health interventions, the challenges highlighted above call for more flexible and agile evaluation methodologies, such as factorial trials, microrandomised trials (MRTs), N-of-1 trials, and simulation-based evaluations. Factorial RCTs are a study design that can be used to evaluate more than one intervention in the same trial.[207] Factorial designs are still underused in mHealth evaluation, but there are a few instances of their adoption to test, for instance, the efficacy of an mHealth app to promote physical activity and fruit and vegetable intake.[208] They have also been employed to test the efficacy of an mHealth intervention to control blood pressure in hypertensive patients,[209] and of a remote intervention to deliver lifestyle treatment for obesity.[210] Despite being a good design to assess behavioural change interventions, factorial trials do not allow researchers to determine the most effective time to deliver intervention options and do not allow measurement of what time-varying factors affect the effectiveness of intervention components.[211] To overcome the limitations of factorial designs, specifically in the context of JITAI development, Klasnjia et al. proposed MRTs.[212] In MRTs, individuals are randomised to different intervention options at multiple time points and therefore allow assessment of the causal effects of each intervention component on the proximal, i.e., short-term, outcomes. Additionally, MRT data allow researchers to examine how the intervention effects change over time. MRTs have been used to evaluate the efficacy of a variety of mHealth solutions, for example aimed at increasing physical activity,[213,214] managing depression and anxiety during the Covid-19 pandemic,[215] helping medical interns cope with stress,[216] and reducing substance abuse.[217] Nonetheless, MRTs still present challenges, including sample size calculations, variable selection, appropriately dealing with missing data, and analysis of non-continuous proximal outcomes.[218]

Other study methods are available to build and evaluate digital behavioural interventions, such as sequential multiple assignment randomised trials (SMARTs) and multiphase optimisation strategy (MOST) described by Collins et al.[219] MOST, presented as an alternative way of building, optimising, and evaluating mHealth interventions, consists of a screening phase to identify effective intervention components, a refining phase to fine-tune these components and determine their optimal levels, and a confirming phase to evaluate the optimised intervention through a standard randomised controlled trial. SMART trials are randomised trials specifically aimed at evaluating adaptive interventions, in which participants are randomised to a new intervention option based on the response to the previous intervention component. SMART trials are increasingly being used to evaluate digital behaviour change interventions, with examples around physical activity,[220] suicidal behaviour,[221] and stress management.[222] Additionally, the SMART design has been paired with computer simulations to identify optimal telemedicine strategies to improve insulin therapy.[223] In particular, simulation-based research has been advocated as an innovative approach with the potential to bridge the evidence gap in digital health by generating more timely evidence at a lower cost.[224]

In summary, while traditional RCTs are crucial for evaluating health interventions, the complexity and rapid evolution of mHealth solutions require more flexible and agile evaluation methodologies. It is essential to adopt new approaches that can address the unique challenges of mHealth interventions, such as usability, engagement, and long-term effectiveness. Employing innovative evaluation techniques allows the development of digital behaviour change solutions that are both effective and adaptable to the diverse needs of users, without the need to wait for RCTs to be completed. Even though RCTs remain the gold standard to prove intervention effectiveness, more agile study designs can provide initial evidence, as well as optimising the development process. Embracing these methods is crucial for advancing the field and maximising the public health impact of mHealth interventions.

CASE STUDY: A PERSONALISED RECOMMENDATION SYSTEM FOR BEHAVIOUR CHANGE IN TYPE 2 DIABETES

As discussed in Chapter 1, T2D is a highly prevalent chronic condition linked to insulin resistance. Obesity, a diet high in carbohydrates, and inactivity are important risk factors and, as such, T2D is initially managed through lifestyle changes like physical activity (PA) promotion and sedentary behaviour (SB) reduction, which have been shown to improve glucose control.[225] To promote a healthier lifestyle effectively, evidence-backed and patient-centred behaviour change strategies are recommended. In particular, technology-enabled interventions can be especially useful in promoting PA and encouraging sedentary time reduction, thus improving glucose control in T2D patients.[226]

mHealth technology is regarded as a technology able to improve health outcomes effectively and at scale due to its widespread appeal, accessibility, and ability to reach large populations at a low cost. Several mHealth interventions have been developed to promote healthy behaviours in people with T2D, with many aimed at improving physical activity and dietary habits.[227,228,229] Despite the widespread availability of advanced technologies, including AI, the vast majority of mHealth interventions for diabetes pair data collected through wrist-worn or smartphone accelerometers with simple data processing techniques. Additionally, these mHealth interventions rarely utilise a holistic approach, which involves grounding the intervention on behavioural theory and personalisation—techniques that have been positively correlated with successful behaviour changes.[230] Where personalisation is used, it is limited to content personalisation, with other personalisable features such as timing, delivery channel, and frequency not being used.[231] Interestingly, while personalisation is effective in changing lifestyle behaviours, the mode of data collection and the source of data used in the personalisation greatly affect the effectiveness of the intervention—specifically, interventions using system-captured data have been associated with higher effectiveness compared to those using user-reported data.[232]

As part of this case study, a rapid systematic review of the literature around mHealth technologies to improve PA, SB, and diet in people with T2D was conducted. The systematic search retrieved 23 studies assessing the efficacy of mHealth interventions. All the mHealth interventions examined employed simple digital techniques, with the most common being SMS, web, and smartphone apps. Rarely did the interventions integrate different data sources—the only ones used are accelerometers and pedometers—and all required manual input from the user. Excluding personalisation achieved through one-on-one sessions with coaches or clinicians, the only personalisation performed by the interventions entailed tailoring the messages sent to users, either through a rule-based algorithm[233] or with reinforcement learning.[234] Only 40% of the studies rooted the mHealth in behavioural theory, and the most common behaviour change techniques were goal setting, self-monitoring, and feedback. Fifteen out of 22 studies (68%) found that the mHealth intervention was effective in either changing behaviours or improving health outcomes such as BMI, waist circumference (WC), and HbA1c. Out of the seven studies that did not find improvements following an mHealth intervention, only two had used behavioural theory (29%). On the other hand, among the studies that did find improvement, 53% used behavioural theory.

The first thing to note from this review is that less than half of the studies included studies selected in the review grounded the intervention in behavioural theory, a design practice that is positively correlated with higher intervention efficacy.[235] This is in line with other reviews highlighting a lack of theoretical content in mHealth applications for physical activity and dietary behaviours.[236,237] Additionally, while personalisation is effective in changing lifestyle behaviours,[238] only a third of the studies employed some sort of personalisation, and only two used a form of automated personalisation. This is in line with reviews highlighting the use of simple forms of personalisation in interventions to increase physical activity.[239]

The aim of this case study is to design and test a new knowledge-based system for personalised lifestyle change. Given the need to use behavioural theories to design interventions in order to achieve higher efficacy,[240] the high-level structure of the system was designed following the COM-B model and the BCW.[241] Additionally, the proposed system, whose high-level diagram is shown in Figure 3.1, enables the behaviour change process by using several layers of personalisation. Firstly, it classifies the user into one of the four diabetes clusters described in the case study in Chapter 1, which are characterised by different cardiometabolic profiles and different associations between physical behaviours and CVD. Secondly, it recommends activity levels based on an individual's genotype and past activity history. Finally, it presents the user with their risk of developing CVD based on their current activity levels. Another layer of personalisation is achieved by including leptin as a feedback biosignal to inform on the efficacy of the intervention towards weight loss. In fact, leptin is a biomarker for obesity management therapies as leptin resistance is related to obesity[242] and leptin levels vary with dieting regimens.[243] No previous system or intervention has used blood chemistry as a feedback signal for behavioural change interventions. Biomarkers have previously been used to assess the efficacy of interventions, but never as an integral part of the recommendation systems. Additionally, the proposed system includes SB alongside PA, which previous works have overlooked, despite it being paramount to improving glucose control in T2D populations. Finally, the system is simulated on data from the UK Biobank, a large-scale clinical database with biomedical data on more than 500,000 participants.

The main components forming the proposed system, shown in Figure 3.1, are:

1. A classifier to categorise new patients in clusters of T2D.
2. The optimised recommendation algorithm, which uses past activity levels, leptin levels, and a polygenic risk score (PRS) to issue the personalised recommendation.
3. A survival model that calculates the risk of developing CVD based on the patient's activity levels.

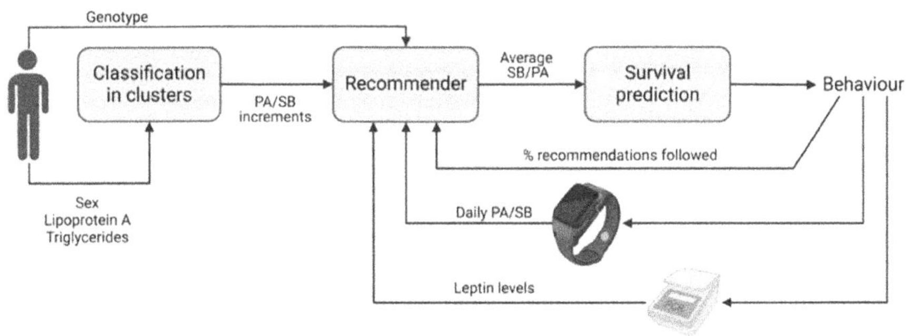

FIGURE 3.1 High-level diagram of the proposed system. Created in BioRender. Cavallo, F. (2024) BioRender.com/o05f919. Reprinted with permission from Cavallo F. R., Toumazou C. (2023) "Personalised lifestyle recommendations for type 2 diabetes: Design and simulation of a recommender system on UK Biobank Data." *PLOS Digit Health* 2(8): e0000333. https://doi.org/10.1371/journal.pdig.0000333

A detailed description of the system design and implementation is provided in the following sections.

Methods: Data source

The system was trained and simulated using data from the UK Biobank. The UK Biobank[244] is a database of over 500,000 adults aged 37–73 recruited in the UK between 2006 and 2010. At recruitment, participants completed a questionnaire and verbal interview about sociodemographic factors, family history, early life exposures, psychosocial factors, environmental factors, lifestyle, and health status. At the same baseline visit, they had physical measures taken and blood, urine, and saliva samples collected. Physical activity and sedentary behaviour data were measured through accelerometer for 100,000 participants for seven days between 2013 and 2015. Ethics approval for the UK Biobank study was obtained from the North West Centre for Research Ethics Committee (REC reference: 21/NW/0157). Informed consent for the UK Biobank study was obtained from participants during the baseline assessment. The work described here was approved by the UK Biobank under application ID 56132. All analyses were performed in accordance with the relevant guidelines and regulations.

In this study, only participants with a self-reported T2D diagnosis, verified by a nurse, were selected and included in the analyses. Due to participants' missing data, the sample sizes vary depending on the analysis (see Table 3.1).

TABLE 3.1
Sample sizes used in the analyses

Analysis type	Description	n
Regression	Participants with diabetes, no missing data in the 16 biomarkers selected, and accelerometer data.	1,805
Regression sensitivity analyses	Participants with diabetes, no missing data in the 16 biomarkers selected, accelerometer data, and no missing data in all confounders (age, waist circumference, income, smoking status).	1,615
Survival analysis	Participants with diabetes, no missing data in the 16 biomarkers selected accelerometer data, and older than 68 years.	1,215
Classification into clusters	Participants with diabetes and older than 68 years (training set), participants with diabetes and younger than 68 years (test set).	24,600, 348
Recommendation system simulation	Participants with diabetes, accelerometer data, younger than 68 years, and with genetic data.	290
Recommender system effect at follow-up	Participants with diabetes, younger than 68 years, with genetic data, and with accelerometer data measures both at baseline and follow-up.	4

METHODS: DESIGN OF THE SYSTEM AS A THEORY-BASED BEHAVIOURAL INTERVENTION

The system was designed following the COM-B model and the BCW. The BCW provides a framework to develop complex interventions that is comprehensive, coherent, and linked to a behavioural model, the COM-B model.

To design the system, the steps needed to develop an intervention according to the BCW were followed, namely:

1. Specifying the target behaviour.
2. Identifying theoretical domains that explain the behaviour.
3. Identifying how to target the theoretical domains.
4. Selecting behaviour change techniques.

Firstly, the target behaviour (i.e., increasing PA and reducing SB) was specified and the associated factors through a literature review. Ovid and Embase databases were systematically searched to include studies identifying barriers and facilitators to the target behaviours, specifically in T2D patients. Five studies were selected (Avery et al. 2015[245] Casey et al. 2010,[246] Cradock et al. 2021,[247] Erickson 2013,[248] Swoboda et al. 2017[249]) that identified factors associated with the target behaviours pertaining to all COM-B model domains (capability, opportunity, and motivation). Common barriers to increasing PA and reducing SB include physical issues and limitations, mental health challenges, lack of motivation, and social support. Environmental factors such as weather, work schedules, time constraints, and availability of suitable environment also represent important barriers to behaviour change. On the facilitator side, techniques such as goal setting, regular supervision and support, and detailed planning are effective in encouraging behaviour change and increasing self-efficacy, which in turn is reflected in sustained motivation to continue engaging in the target behaviour. Motivation is also increased by the knowledge that lifestyle changes can delay the use of medications to manage T2D, as well as by feedback on the positive effects of PA on glycaemic control and tangible health benefits.[250]

All factors associated with the target behaviours that were identified by the selected papers are presented in Table 3.2, including those that cannot be addressed by an mHealth intervention (such as environmental factors), in order to have a more complete picture of the elements influencing physical behaviours in people with T2D. Secondly, the identified factors were mapped to the COM-B model. And lastly, suitable behaviour change techniques were identified through the help of the TDF)[251] and mapped to system features. The design process is shown in Table 3.2.

According to the COM-B model, an intervention must match the patient's abilities to be effective and sustained. Challenges recommended by the system have to align with the patient's skills and constantly adjust as these change over time. Therefore, in the proposed system, the recommendations are initialised to the patient's baseline activity and optimised over time depending on the probability of compliance and leptin levels. The probability of compliance with the recommendations is paramount

TABLE 3.2
Implementations of the BCW for the design of a knowledge-based system for personalised PA and SB recommendations

Factors associated to behaviours	COM-B domain	TDF domains chosen	System feature
• Physical health and limitations • Lacking energy • Increased awareness of own skills	Physical capability	Skills	Progressive increase from baseline *
• Self-efficacy • Mental health • Time management • Initial cognitive demand • Planning • Effort to change T2D management • Knowledge about PA for T2D management	Psychological capability	• Knowledge • Skills • Self-monitoring • Action planning	Self-monitoring trends *
• Social support • Supervision and encouragement	Social opportunity	Social comparison	Score boards
• Focus on other tasks • Time constraints and other commitments • Weather conditions • Availability of suitable environment • Cost of equipment	Physical opportunity	Barriers and facilitators	Notifications and reminders
Encouragement	Automatic motivation	Reinforcement	Goal achievement and reinforcement
• Individualised exercise • Feedback on impact of PA on health • Knowledge of benefits • Lack of motivation • Motivated by tangible benefits • Motivated by not needing medication	Reflective motivation	• Beliefs about capabilities • Beliefs about consequences • Goals	• Rewards • Personalised PA/ SB levels * • Time-to-onset lost or saved * • Goal setting

Reprinted with permission from Cavallo F. R., Toumazou C. (2023) "Personalised lifestyle recommendations for type 2 diabetes: Design and simulation of a recommender system on UK Biobank Data." *PLOS Digit Health* 2(8): e0000333. https://doi.org/10.1371/journal.pdig.0000333
Entries marked with (*) have been implemented in this work.

to prevent feelings of failure in the patient, which can hinder engagement with the intervention. Moreover, PA and SB trends enable self-monitoring and self-regulation, which increase motivation. Finally, rewards are embedded in the system as a time-to-onset of comorbidities gained or lost in response to their new activity behaviour.

Methods: Classification into diabetes clusters

The first component of the proposed system is the classification of patients into the diabetes clusters.[252] To select the best classifier, three models—K-nearest neighbours (KNN), support vector classifier (SVC), and random forest (RF)—were trained and validated on the clustering set ($n = 26{,}400$). The test set was selected as the set of participants under 68 years old ($n = 348$) later used in the prediction of survival times, as this set is used to simulate the system and thus should not be used to train the classifier.

For each of the three models, features were standardised and selected through forward sequential feature selection in *mlxtend*,[253] using balanced accuracy as the selection metric. The model hyperparameters were tuned with randomised search cross-validation, using the *scikit-learn* RandomizedSearchCV function which implements a fit and score method,[254] with balanced accuracy being selected as the scoring method. The tuned models with the optimal number of features were then validated on the test set.

Methods: Optimised recommendation module

The main component of the system is the recommendation module that issues optimised recommendations to improve metabolic health while keeping the probability of compliance high. The core inputs for the recommendation module are leptin levels, probability of compliance, and upper and lower limits of PA and SB, expressed as minutes per day.

Leptin levels vary in relation to dieting therapies and thus are a real-time physiological signal providing feedback on the efficacy of the intervention. During dietary interventions, leptin levels change at a greater rate than weight and can predict weight loss two weeks in advance.[255] Therefore, monitoring leptin can be more meaningful than monitoring weight, and leptin can be employed as feedback to adjust recommendations if future weight gain is predicted. Monitoring leptin levels is also relevant for chronic disease management, as hyperleptinemia is correlated with an increased risk of obesity and cardiovascular outcomes in T2D patients.[256]

The probability of compliance (Equation (3.1)) is related to the historical levels of activity and, therefore, encodes the skill level of the patients. In order to keep the compliance high, the recommendations are adjusted based on the proportion of recommendations followed in the past and the user's past activity levels. To include the effect of genetics on motivation, the recommendation score is also weighted by a polygenic risk score. The polygenic risk score is calculated using effect sizes reported by Doherty et al.[257] that quantify the association between selected single nucleotide polymorphisms (SNP) and PA or SB. Information on the selected SNP is given in Table 3.3.

The probability of compliance is calculated as follows:

$$P(c) = \frac{\sum_{i=1}^{R} f_i}{R} \times \left(1 - \left|A_{\mathrm{r}} - A_{\mathrm{p}}\right|\right) + PRS \tag{3.1}$$

TABLE 3.3
Genetic markers used in the polygenic risk score calculation

Trait	SNP	Chromosome	Nearest gene	Effect allele	Alternative allele	AF
PA	rs6775319	3	SATB1-AS1	A	T	0.271
PA	rs6895232	5	LINC01470	T	A	0.663
PA	rs564819152	10	SKIDA1	A	G	0.679
PA	rs2696625	17	KANSL1-AS1	A	G	0.77
PA	rs59499656	18	SYT4	A	T	0.655
SB	rs61776614	1	FAAP20	C	T	0.925
SB	rs1858242	3	LOC105377146	A	G	0.259
SB	rs26579	5	MEF2C-AS2	G	C	0.415
SB	rs25981	5	EFNA5	G	C	0.531
SB	rs6870096	5	NMUR2	C	G	0.321
SB	rs34858520	7	CALN1	A	G	0.558

SNP: single nucleotide polymorphism, AF: allele frequency

$$PRS = \sum_{i=1}^{N} g_i \cdot e_i$$

$$f_i = \begin{cases} 1, \textit{ if recommendation followed} \\ 0, \qquad \textit{otherwise} \end{cases} \tag{3.2}$$

where R is the number of recommendations issued, PRS is the polygenic risk score, g_i is the genotype call, and e_i is the effect size corresponds to SNP i. The factor $\alpha = 1 - |A_r - A_p|$ represents the influence of past activity on the probability of compliance, where A_r is the new recommended activity and A_p is the predicted activity estimated through fitting a logistic function on the historical activity levels. To ensure that $0 \le \alpha \le 1$, the recommended and predicted activity are calculated as daily proportions. The decision-based algorithm to determine the optimised recommendation is shown in Figure 3.2.

Methods: Calculation of cluster-specific activity increments

To determine the daily change in PA and SB to apply for each cluster, stratified regression analyses were performed for each diabetes subtype. The reader is referred to Chapter 1 for a detailed description of the clustering analysis of T2D patients in the UK Biobank.

To calculate the cluster-specific activity increments necessary to lower the risk of CVD, logistic regression was conducted with Python's *statsmodels*. After removing

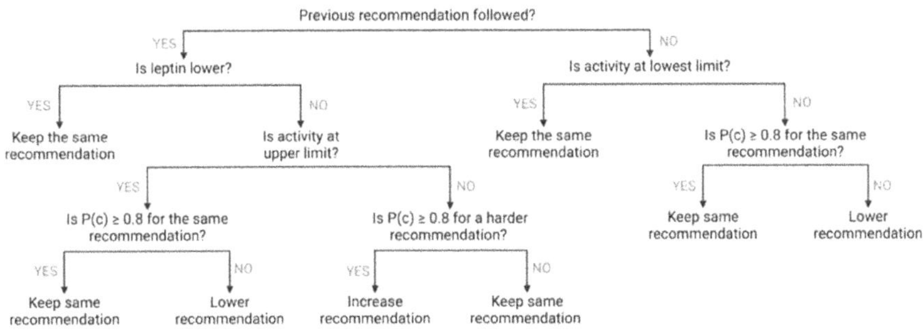

FIGURE 3.2 Diagram of the algorithm to generate the optimised recommendations using historical activity levels and the probability of compliance to the updated recommendation. Reprinted with permission from Cavallo F. R., Toumazou C. (2023) "Personalised lifestyle recommendations for type 2 diabetes: Design and simulation of a recommender system on UK Biobank Data." *PLOS Digit Health* 2(8): e0000333. https://doi.org/10.1371/journal.pdig.0000333

participants with missing accelerometer data, or without at least three days of data and data in each one-hour period of the 24-hour cycle (scattered over multiple days), n = 1,805 participants were included in the regression analysis.

Firstly, the raw accelerometer data were classified into SB, light PA, walking, moderate-to-vigorous PA (MVPA), and sleep with balanced random forests and Hidden Markov models using the accelerometer package by Willets et al.[258] Despite the participants being instructed to wear the accelerometer all the time, non-wear segments were present and were imputed using the average of similar time-of-day data points taken from different days of the week. Secondly, the day was partitioned into time spent in overall PA (including time spent in light PA, walking, and MVPA), SB, and sleep, and we included the accelerometer in the regression model with a compositional approach.[259] In a compositional framework, time spent in the three different activities is considered as a relative proportion of the overall time budget (24 hours), and thus conventional statistical methods cannot be employed with compositional data[260] due to the implied collinearity between the activities. Therefore, the isometric log-ratio (ILR) transformation was used. A detailed explanation of the ILR transformation can be found in Chapter 2. The ILR regression coefficients were also used to estimate the minutes of activity required to reduce the log odds of CVD.

The ILR coefficients obtained with logistic regression can be used to estimate the minutes of activity required to reduce the log(odds) of CVD, as Dumuid et al.[261] proved that for a three-activity composition, the following holds:

$$\Delta CVD(k) = \beta_1 \sqrt{\frac{3}{4}} \ln k \tag{3.3}$$

$$k = \frac{1+r}{1-s}, \qquad s = r \cdot \frac{\alpha_i}{\alpha_{24h-i}} \tag{3.4}$$

with β_1 being the first regression coefficient, k being change in daily activity composition, r being the change in primary activity α_i compared to the sample mean, and s being the corresponding change in the remaining activity components α_{24h-i}.

The effect of common confounders, including age, WC, alcohol intake, income, and smoking status as covariates, was examined and the final model was selected with backward selection—only covariates reaching significance (p-value < 0.05) were retained in the final model. Since only WC, smoking status, and age were significant covariates, participants with complete data for these variables were included in the analysis (n = 1,805). Sensitivity analyses were done on a subset of participants with no missing data for all covariates (including alcohol intake and income) with n = 1,615.

Methods: Survival analysis

The effects of PA and SB on CVD onset were assessed in the different clusters with the Kaplan–Meyer Estimate. In order to predict survival time from PA and SB levels simultaneously, activity data were not transformed with ILR transformation and were used as minutes/day. The time between birth and onset of CVD was used as the time-to-event of interest, with the event being diagnosis of CVD. The observations were censored at 68 years, and subjects who had not reached this age at the baseline visit were removed, leaving a sample size of n = 1,215.

Survival regression was further conducted to assess the effect of relevant risk factors on survival time. Smoking, sex, age, triglycerides, diastolic blood pressure (DBP), total cholesterol, LDL-cholesterol, and WC were included as covariates in the models, as they are the most common predictors for CVD.[262] After examination of the partial residual plots, a non-linear term (age^2_{T2D}) and two interaction terms (PA × age and PA × DBP) were included. All features were scaled to the [0, 1] range. Four models were compared with k-fold cross-validation (k = 10, 100 iterations): Cox's proportional hazards, Weibull AFT, Log-logistic and Log-normal AFT models. The analysis was done with the Python *lifelines* package.[263] The selected model is then used in the recommender system to predict the survival time for each user and to communicate the risk of developing CVD in response to one's levels of PA and SB.

Methods: Simulations

To simulate the effects of the recommender system without real-life users, we simulated physical activity based on whether the proposed recommendation was followed or not, which was modelled through different probability distributions. If the recommendation was followed, the PA (or SB) level is set equal to the recommendation; otherwise, it is set equal to the level recorded (or simulated) on the previous day. PA and SB levels are reported as a portion of the day. We used the data presented by Finkelstein et al.[264] and Meyer et al.[265] to model the probability of following a recommendation over time. The effectiveness of the system in inducing behaviour change was simulated on the young dataset with no missing genetic data (n = 290).

The proposed system was simulated with the aforementioned models for a duration of 365 days.

As leptin levels were not available in the UK Biobank, we assigned baseline leptin levels using the regression coefficients estimated by Zuo et al.[266] to predict leptin from gender, age, BMI, WC, arm fat mass, energy intake, PA, and smoking. Missing data were replaced by the sample mean. However, for energy intake, there was a significant amount of missing data, and values were replaced through multivariate sample imputation. Leptin was updated based on the results of a systematic review by Fedewa et al.[267] that found that engaging in chronic exercise (⋝2 weeks) is associated with a reduction of leptin with an effect size of 0.24. Consequently, leptin is reduced by the effect size if the recommendations are followed for 2 weeks; otherwise, it is kept unchanged.

Results: Cluster analysis

Given the similar cluster characteristics (see Chapter 2), males and females were combined for regression analysis. After backward feature selection, the final models for the clusters are:

$$CVD = z_1 + z_2 + WC + Age \qquad \text{Cluster 1} \qquad (3.5)$$

$$CVD = z_1 + z_2 \qquad \text{Clusters 2 and 4} \qquad (3.6)$$

$$CVD = z_1 + z_2 + WC + Smoking \qquad \text{Cluster 3} \qquad (3.7)$$

where z_1 and z_2 are the first and second coefficients of the ILR transformation. The odds of CVD and mean time spent in PA and SB for each cluster are presented in Table 3.4. The time steps required to reduce the risk of CVD by 1% for each cluster are presented in Table 3.5. Some of the steps in Table 3.5 were adjusted, given the wrong step direction due to p-value > 0.05. Specifically, the PA time steps for Clusters 2 and 4 are replaced by the time step for the whole sample (20 minutes/day). The SB time step for Cluster 2 is set to 18 minutes/day, given the similarity in SB levels with Clusters 1 and 3.

The effect of PA and SB on the CVD survival function was assessed. Figure 3.3 shows the Kaplan–Meier survival functions estimated on censored subjects (n = 1,215), where the biggest difference is observed between Clusters 1 and 4. This result is expected, since Cluster 4 presents a worse cardiometabolic profile than Cluster 1 and thus, is more prone to CVD development. Figure 3.4 shows the effects of changing SB and PA on Cluster 2 as a representative example; the effect of PA is visibly pronounced, while the effects of varying sedentary time are more subtle.

For survival regression, the Cox's model was selected as it obtained the highest concordance score (0.74 for Cox, Weibull, and Log logit, 0.73 for Log normal), which

TABLE 3.4
Odds of CVD and geometric mean of time spend in physical activity by cluster

Cluster	n_{ill}	$n_{healthy}$	PA (mins/day)	SB (mins/day)	CVD odds
Sample	1,264	351	243	509	3.6
1	391	138	262	487	2.8
2	326	84	241	514	3.9
3	224	61	233	522	3.7
4	323	68	229	528	4.8

Reprinted with permission from Cavallo F. R., Toumazou C. (2023) "Personalised lifestyle recommendations for type 2 diabetes: Design and simulation of a recommender system on UK Biobank Data." *PLOS Digit Health* 2(8): e0000333. https://doi.org/10.1371/journal.pdig.0000333

TABLE 3.5
Associations between CVD and compositional measures of physical activity and conversion from compositional coefficients to time (minutes/day) of physical activity and sedentary behaviour required to reduce the risk of CVD by 1%

PA	n	log-odds (p-value)	Step (mins)
Sample	1,805	−0.41(0.009)	20
Cluster1	602	−0.83 (0.001)	8
Cluster2	445	0.06 (0.84)	−108
Cluster3	432	−1.45 (0.0003)	6
Cluster4	326	0.2 (0.55)	−47
SB	n	**log-odds (p-value)**	**Step (mins)**
Sample	1805	0.17 (0.44)	−72
Cluster1	602	0.51 (0.15)	−19
Cluster2	445	−0.84 (0.07)	16
Cluster3	432	0.74 (0.13)	−17
Cluster4	326	0.11 (0.82)	−141

Reprinted with permission from Cavallo F. R., Toumazou C. (2023) "Personalised lifestyle recommendations for type 2 diabetes: Design and simulation of a recommender system on UK Biobank Data." *PLOS Digit Health* 2(8): e0000333. https://doi.org/10.1371/journal.pdig.0000333

The log-odds correspond to change in CVD log-odds for increases in time spent in the given behaviour and corresponding decrease in the others

measures the accuracy of the ranking of predicted times. The estimated Cox's model was used to predict the cluster-dependent survival times for subjects younger than 68 years. Survival times are calculated as the month at which the survival function crosses a probability $p = 0.5$ and are shown in Table 3.6.

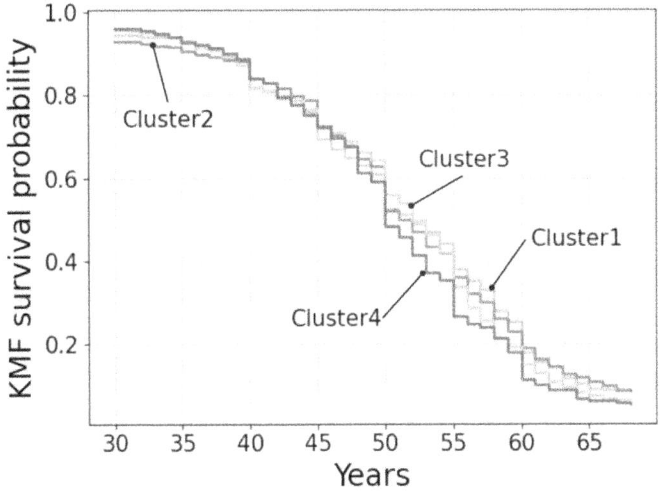

FIGURE 3.3 Kaplan–Meier survival probability for the four clusters. Reprinted with permission from Cavallo F. R., Toumazou C. (2023) "Personalised lifestyle recommendations for type 2 diabetes: Design and simulation of a recommender system on UK Biobank Data." *PLOS Digit Health* 2(8): e0000333. https://doi.org/10.1371/journal.pdig.0000333

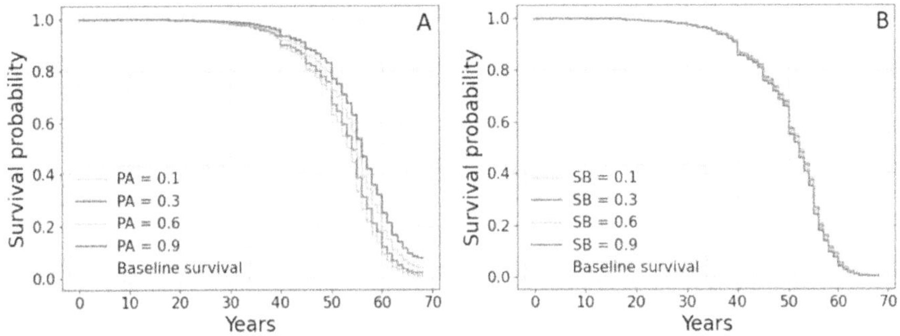

FIGURE 3.4 Partial effects of SB (A) and PA (B) on the survival function. Reprinted with permission from Cavallo F. R., Toumazou C. (2023) "Personalised lifestyle recommendations for type 2 diabetes: Design and simulation of a recommender system on UK Biobank Data." *PLOS Digit Health* 2(8): e0000333. https://doi.org/10.1371/journal.pdig.0000333

Results: Recommender system classifier selection

The first component of the proposed system is the classification of patients into the diabetes clusters. Based on the best accuracy scores, three features were selected for KNN and SVC (sex, lipoprotein A, and triglycerides), while four were selected for RF (sex, pulse rate, LDL-cholesterol, and triglycerides). While the three classifiers achieved comparable performance, the support vector classifier was chosen as the preferred classifier as it uses fewer features than the RF and reached a better accuracy than the KNN. The performance for each classifier tested is shown in Table 3.7.

TABLE 3.6
Cox's proportional hazards model prediction on censored subjects

Cluster	n_{train}	n_{young}	Predicted survival for *young* set
1	365	143	19 (11)
2	306	80	23 (12)
3	314	64	20 (9)
4	230	61	21 (13)

Reprinted with permission from Cavallo F. R., Toumazou C. (2023) "Personalised lifestyle recommendations for type 2 diabetes: Design and simulation of a recommender system on UK Biobank Data." *PLOS Digit Health* 2(8): e0000333. https://doi.org/10 1371/journal.pdig.0000333
Data presented as mean (standard deviation).

TABLE 3.7
Comparison of classifiers performance and number of features selected through sequential feature selection

Classifier	Number of features	Training accuracy (n = 13,423)	Test accuracy (n = 348)
KNN	3	0.98	0.99
SVC	3	0.99	0.994
RF	4	1	1

Reprinted with permission from Cavallo F. R., Toumazou C. (2023) "Personalised lifestyle recommendations for type 2 diabetes: Design and simulation of a recommender system on UK Biobank Data." *PLOS Digit Health* 2(8): e0000333. https://doi .org/10.1371/journal.pdig.0000333
KNN: K-nearest neighbours, SVC: support vector classifier, RF: random forest.

Results: Simulated effect of system on behaviour change

The proposed system was simulated with the aforementioned models for a duration of 365 days. The probability of following the recommendations on any given day was sampled from a Bernoulli distribution, and the parameter p was modelled with three different functions that reflect the time-dependent nature of adherence to human interventions. The models were estimated from data reported by Finkelstein et al.[268] (model in Equation (3.8), shown in Figure 3.5A) who measured the percentage of participants who wore the activity tracker for at least one day in a week, and by Meyer et al.[269] (model in Equation (3.9), shown in Figure 3.5B), who recorded the average number of days in a week that participants wore the tracker. The third model was time-invariant, with constant p over the duration of the intervention (Equation (3.10)). The equations describing the models for simulating adherence are the following, where t represents time and where the

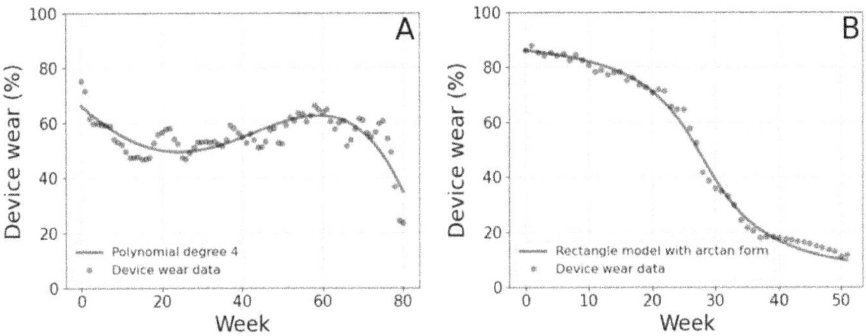

FIGURE 3.5 Progression of accelerometer device wear over time as reported by Finkelstein et al. (A) and Meyer et al. (B). The data were used to fit the models for adherence probability described in Equations (3.8) and (3.9), respectively. Reprinted with permission from Cavallo F. R., Toumazou C. (2023) "Personalised lifestyle recommendations for type 2 diabetes: Design and simulation of a recommender system on UK Biobank Data." *PLOS Digit Health* 2(8): e0000333. https://doi.org/10.1371/journal.pdig.0000333

other parameters for Equations (3.8) and (3.9) are estimated by fitting the data by Finkelstein et al.[270] and Meyer et al.:[271]

$$P(X=1) = p(t; A, \mu_1, \mu_2, \sigma_1, \sigma_2) \tag{3.8}$$

$$= \frac{A}{\pi}\left[\arctan\left(\frac{t-\mu_1}{\sigma_1}\right) + \arctan\left(\frac{t-\mu_2}{\sigma_2}\right)\right]$$

$$P(X=1) = p(t; \beta_1, \beta_3, \beta_3, \beta_4, \beta_5) \tag{3.9}$$

$$= \beta_1 t^4 + \beta_2 t^3 + \beta_3 t^2 + \beta_4 t + \beta_5$$

$$P(X=1) = p \quad \text{with } p = (0,1) \tag{3.10}$$

The resulting activity and adherence profiles for the *young* dataset are presented in Figure 3.6. Within the *young* dataset, four participants had follow-up accelerometer data available, collected approximately 1.5 years after the baseline measurements. Consequently, it was possible to compare the activity at follow-up with the simulated activity at a time matching the time difference between baseline and follow-up measurements. Figure 3.7 shows that, although only a constantly high adherence (p = 0.8) to the intervention would bring significant improvements in activity compared

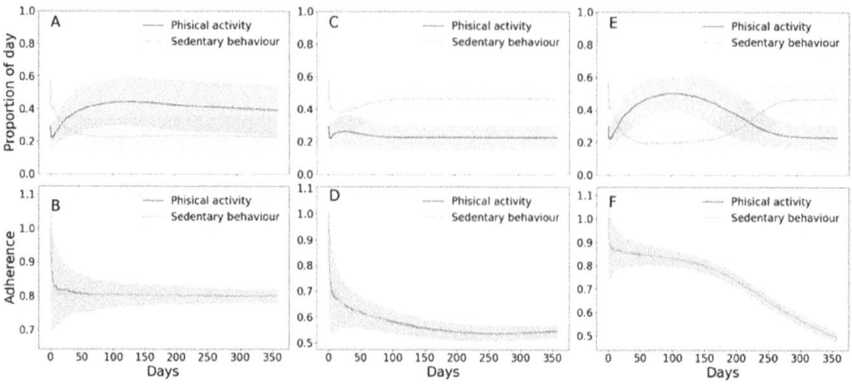

FIGURE 3.6 Simulated activity and adherence (i.e., proportion of recommendations followed) with time-varying probabilities. The probabilities are modelled according to (3.8) with $p = 0.8$ (A, B), Equation (3.9) (C, D) and equation (3.10) (E, F). The top graphs (A, C, E) show the proportion of the day spent in physical activity and sedentary behaviour across the simulated intervention. The bottom graphs (B, D, F) show the simulated adherence to the intervention as a function of days. Data presented as mean and standard deviation. Reprinted with permission from Cavallo F. R., Toumazou C. (2023) "Personalised lifestyle recommendations for type 2 diabetes: Design and simulation of a recommender system on UK Biobank Data." *PLOS Digit Health* 2(8): e0000333. https://doi.org/10.1371/journal.pdig.0000333

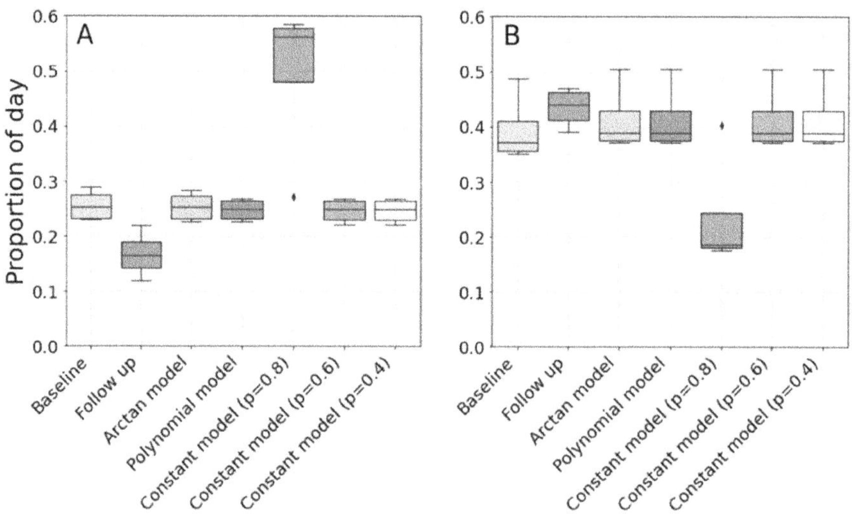

FIGURE 3.7 Comparison between simulated physical activity (A) and sedentary behaviour (B) after one year of personalised intervention and real-life measured activity at baseline and follow-up. Reprinted with permission from Cavallo F. R., Toumazou C. (2023) "Personalised lifestyle recommendations for type 2 diabetes: Design and simulation of a recommender system on UK Biobank Data." *PLOS Digit Health* 2(8): e0000333. https://doi.org/10.1371/journal.pdig.0000333

to baseline, all adherence models achieved improvements compared to activity measured at follow-up. For PA, such a difference was statistically significant for all models evaluated (p-value < 0.05, tested with a paired t-test).

Discussion

This case study presented a new knowledge-based system for personalised activity recommendations. The system aims to increase PA and reduce SB in people with T2D by providing recommendations with a high chance of compliance based on the patient's T2D cluster, previous activity, genetic risk, and leptin levels. The system was trained and simulated using data from the UK Biobank, and it was demonstrated that the system may help increase levels of activity. In fact, simulating the system with different probabilities of adherence showed improved physical activity compared to real-life activity measured after 1.5 years from baseline. Despite testing the system with computer simulations, the behavioural response to the system was modelled using real data collected on human participants by two studies in which the weekly wear of an activity tracker was recorded for 52 weeks. While these data are suitable for initial assessment, implementation of the system as a real-life intervention and user testing are needed to reliably assess the effectiveness of the system.

The novelty of the presented work lies in the adoption of a holistic approach to behaviour change interventions: in fact, a wide array of data was used—including biomarkers, genotype, leptin levels, and past activity—to achieve highly personalised recommendations. Another novel concept in this work is the use of genetic information to optimise the personalised recommendations. Genetic risk communication has been previously used for lifestyle behaviour change interventions but has been found to have a very small effect on reducing unhealthy behaviours such as smoking, physical activity, and nutrition.[272] However, to the authors' knowledge, no intervention has been developed that uses genetic markers associated with the target behaviour. The inclusion of such markers may contribute to improving lifestyle behaviours because, in the present system, genetic information is related to the capability of the subject. On the other hand, previous interventions have mapped genetic information to motivation and specifically to the knowledge of the risks associated with a poor lifestyle. Future research should confirm whether genetic associations can be used to directly inform lifestyle recommendations and achieve a significant change in lifestyle behaviours.

The work in this case study presents several limitations. Firstly, the behavioural analysis is limited as it only employed findings from a literature review, while conducting new qualitative research with the target population could have provided further insights into barriers and facilitators of the target behaviour, specifically in regard to mHealth interventions. Nonetheless, it is possible to improve the behavioural analysis by conducting focus groups or semi-structured interviews with the target population and incorporating the results in future versions of the proposed system.

The system is evaluated through computer simulations rather than through an RCT, which is the gold standard to assess the efficacy of an intervention.[273] A direct consequence is the impossibility of implementing features, such as goal customisation, which require the user's input. Additionally, at this stage, it is not possible to assess the system's user-friendliness and how it would impact the acceptance and efficacy of the intervention. It is also not possible to assess whether communicating the risk of developing CVD is the best way to increase motivation, and other rewards may be more effective. Additionally, several issues can undermine the efficacy of the system in real-life settings, which could not be accounted for in simulations. Firstly, the engagement with the system is currently modelled with time-changing probability distributions, which may be different from the real-world adherence to the system's recommendations. Secondly, simulations cannot capture the individual variation in leptin levels in different individuals and the response to different types of physical activity. Therefore, modelling leptin can introduce errors in the optimisation of recommendations and thus in the efficacy simulations. Thirdly, simulations cannot account for all the factors that may affect users' activity and behaviour and therefore, the system efficacy may be invalidated in free-living settings.

A further limitation of the current system version is the use of overall activity, which combines any activity not classified as sedentary time or sleep. Classifying activities based on type and intensity may be more beneficial for the user as it can provide a more precise indication of which kind of activity one should replace sedentary time with.

The promising results presented in this case study warrant the implementation of the system into a real-life intervention, whose efficacy can be tested with an RCT. To implement the intervention, the system would be integrated with a point-of-care test for leptin.[274,275] Additionally, the recommendations can be augmented with information on activity mode and duration, which are especially important when dealing with cardiometabolic conditions that require special considerations due to the associated risks. Additional gamification features, such as goal setting and scoreboards, can be integrated into the system to make it customisable by the user and increase self-efficacy. As data availability increases and new features are added to the system, data can be represented using knowledge graphs to capture complex relationships and better integrate data from different sources. Specifically, a personal knowledge graph[276] would be used to define structural relationships between the patient and their data, which can then be used to improve recommendations, infer missing data, and predict disease risk.

However, before implementing the system into a real-world intervention, several considerations have to be made. While personalisation promotes acceptance of mHealth services,[277] it also raises privacy and data security concerns. Tangari et al.[278] found that very few mHealth apps respect user privacy—88% of apps found in the major app stores can access and share personal data. As the present system relies on sensitive health data, its implementation must ensure privacy and user data security by using encrypted HTTP traffic and encrypting the data stored on the user's device. Therefore, while mHealth applications often present a lack of data security, ensuring a high level of data protection is achievable in order to optimise the risk-benefit trade-off of mHealth interventions.

Once the initial implementation of the mHealth system is completed according to the best data protection practices, a first round of user testing should be conducted to assess the system's usability, perceived benefits, and prospective acceptability.

Subsequently, the system should be updated to reflect the feedback collected during the user trial. Finally, an RCT would assess the efficacy of the proposed system, which is a first attempt at providing comprehensive and highly personalised activity recommendations by linking sensing data with genetics and daily tracking. If proven effective and acceptable in real-life settings, implementation of such theories could have a major impact on the prevention and management of chronic diseases that are triggered by behavioural practices.

This case study has been previously published in Cavallo F. R., Toumazou C. (2023) "Personalised lifestyle recommendations for type 2 diabetes: Design and simulation of a recommender system on UK Biobank Data." *PLOS Digit Health* 2(8): e0000333. https://doi.org/10.1371/journal.pdig.0000333

ABBREVIATIONS

AI	Artificial intelligence
BCW	Behaviour change wheel
BMI	Body mass index
COM-B	Capability–Opportunity–Motivation Behaviour model
DBP	Diastolic blood pressure
EHR	Electronic health record
EMI	Ecological momentary intervention
EMR	Electronic medical record
HAPA	Health action process approach
HbA1c	Haemoglobin A1c
HBM	Health belief model
HDL	High-density lipoprotein
ILR	Isometric log ratio
IMB	Information Motivation Behavioural Skills model
JITAI	Just-in-time adaptive intervention
KNN	k-nearest neighbours
LDL	Low-density lipoprotein
LLM	Large language model
mHealth	Mobile health
ML	Machine learning
MOST	Multiphase optimisation strategy
MRT	Micro randomised trial
NICE	National Institute of Health and Care Excellence
OECD	Organisation for Economic Cooperation and Development
RCT	Randomised controlled trial
RF	Random forest
RNN	Recurrent neural network
SMART	Sequential multiple assignment randomised trial
SNP	Single nucleotide polymorphism
SVC	Support vector classifier
T2D	Type 2 diabetes

TDF	Theoretical domains framework
TPB	Theory of Planned Behaviour
TTM	Transtheoretical model
WC	Waist circumference
WHO	World Health Organization

NOTES

1. Collaborators, "Global, Regional, and National Comparative Risk Assessment of 79 Behavioural, Environmental and Occupational, and Metabolic Risks or Clusters of Risks, 1990–2015: A Systematic Analysis for the Global Burden of Disease Study 2015."
2. N. Egaddal, E. A. Kramarow, and C. Reuben, "Physical Activity among Adults Aged 18 and Over: United States, 2020," in *NCHS Data Brief, no 443* (Hyattsville, MD: National Center for Health Statistics, 2022).
3. Sport England, "Active Lives Adult Survey November 2021–22 Report," (2023).
4. M. P. Kelly and M. Barker, "Why Is Changing Health-Related Behaviour So Difficult?," *Public Health* 136 (2016): 109-16.
5. Kelly and Barker, "Why Is Changing Health-Related Behaviour So Difficult?"
6. H. Leventhal and L. Cameron, "Behavioral Theories and the Problem of Compliance," *Patient Educ* 10, no. 2 (1987): 117-38.
7. J. D. Fisher and W. A. Fisher, "Changing Aids-Risk Behavior," *Psychol Bull* 111, no. 3 (1992): 455–74.
8. Icek Ajzen, "The Theory of Planned Behavior," *Organ Behav Hum Decis Process* 50, no. 2 (1991): 179–211.
9. S. Michie, M. M. van Stralen, and R. West, "The Behaviour Change Wheel: A New Method for Characterising and Designing Behaviour Change Interventions," *Implement Sci* 6 (2011): 42.
10. F. Monteiro-Guerra et al., "Personalization in Real-Time Physical Activity Coaching Using Mobile Applications: A Scoping Review," *IEEE J Biomed Health Inform* 24, no. 6 (2020): 1738–51.
11. J. Horne et al., "Enhanced Long-Term Dietary Change and Adherence in a Nutrigenomics-Guided Lifestyle Intervention Compared to a Population-Based (Glb/Dpp) Lifestyle Intervention for Weight Management: Results from the Now Randomised Controlled Trial," *BMJ Nutr Prev Health* 3, no. 1 (2020): 49-59.
12. R. P. Hawkins et al., "Understanding Tailoring in Communicating About Health," *Health Educ Res* 23, no. 3 (2008): 454-66.
13. N. Bol et al., "Customization in Mobile Health Apps: Explaining Effects on Physical Activity Intentions by the Need for Autonomy," *Digit Health* 5 (2019): 2055207619888074.
14. A. Bandura, "Self-Efficacy: Toward a Unifying Theory of Behavioral Change," *Adv Behav Res Therapy* 1, no. 4 (1978): 139–61.
15. D. E Nielsen and A. El-Sohemy, "A Randomized Trial of Genetic Information for Personalized Nutrition," *Genes Nutr* 7, no. 4 (2012): 559-66.
16. D. E. Nielsen and A. El-Sohemy, "Disclosure of Genetic Information and Change in Dietary Intake: A Randomized Controlled Trial," *PLoS One* 9, no. 11 (2014): e112665.
17. C. S. Bloss, N. J. Schork, and E. J. Topol, "Effect of Direct-to-Consumer Genomewide Profiling to Assess Disease Risk," *N Engl J Med* 364, no. 6 (2011): 524-34.
18. G. J. Hollands et al., "The Impact of Communicating Genetic Risks of Disease on Risk-Reducing Health Behaviour: Systematic Review with Meta-Analysis," *BMJ* 352 (2016): i1102.

19. T. M. Marteau et al., "Effects of Communicating DNA-Based Disease Risk Estimates on Risk-Reducing Behaviours," *Cochrane Database Syst Rev*, no. 10 (2010): CD007275.
20. U. Görman et al., "Do We Know Enough? A Scientific and Ethical Analysis of the Basis for Genetic-Based Personalized Nutrition," *Genes Nutr* 8 (2013): 373-81.
21. C. B. O'Donovan et al., "Knowing Your Genes: Does This Impact Behaviour Change?," *Proc Nutr Soc* 76, no. 3 (2017): 182-91.
22. S. X. Li et al., "The Effect of Communicating the Genetic Risk of Cardiometabolic Disorders on Motivation and Actual Engagement in Preventative Lifestyle Modification and Clinical Outcome: A Systematic Review and Meta-Analysis of Randomised Controlled Trials," *Br J Nutr* 116, no. 5 (2016): 924-34.
23. H. C. Price, S. J. Griffin, and R. R. Holman, "Impact of Personalized Cardiovascular Disease Risk Estimates on Physical Activity—a Randomized Controlled Trial," *Diabet Med* 28, no. 3 (2011): 363-72.
24. R. Bize et al., "Biomedical Risk Assessment as an Aid for Smoking Cessation," *Cochrane Database Syst Rev* 3, no. 12 (2012): CD0047.
25. A. G. K. Edwards et al., "Personalised Risk Communication for Informed Decision Making About Taking Screening Tests," *Cochrane Database Syst Rev*, no. 2 (2013): CD001865.
26. S.Chaiken and Y. Trope, *Dual-Process Theories in Social Psychology* (Guilford Press, 1999).
27. D. P. French et al., "Can Communicating Personalised Disease Risk Promote Healthy Behaviour Change? A Systematic Review of Systematic Reviews," *Ann Behav Med* 51, no. 5 (2017): 718-29.
28. G. J. Hollands, M. Hankins, and T. M. Marteau, "Visual Feedback of Individuals' Medical Imaging Results for Changing Health Behaviour," *Cochrane Database Syst Rev*, no. 1 (2010): CD007434.
29. H.-L. Hietaranta-Luoma et al., "An Intervention Study of Individual, Apoe Genotype-Based Dietary and Physical-Activity Advice: Impact on Health Behavior," *J Nutrigenet Nutrigenomics* 7, no. 3 (2015): 161-74.
30. Horne et al., "Enhanced Long-Term Dietary Change and Adherence in a Nutrigenomics-Guided Lifestyle Intervention Compared to a Population-Based (Glb/Dpp) Lifestyle Intervention for Weight Management: Results from the Now Randomised Controlled Trial."
31. I. J. Kullo et al., "Incorporating a Genetic Risk Score into Coronary Heart Disease Risk Estimates," *Circulation* 133, no. 12 (2016): 1181-8.
32. J. Horne et al., "A Systematic Review of Genetic Testing and Lifestyle Behaviour Change: Are We Using High-Quality Genetic Interventions and Considering Behaviour Change Theory?," *Lifestyle Genom* 11, no. 1 (2018): 49-63.
33. A. King et al., "The Efficacy of Genotype-Based Dietary or Physical Activity Advice in Changing Behavior to Reduce the Risk of Cardiovascular Disease, Type II Diabetes Mellitus or Obesity: A Systematic Review and Meta-Analysis," *Nutr Rev* 81, no. 10 (2023): 1235-53.
34. Kelly and Barker, "Why Is Changing Health-Related Behaviour So Difficult?."
35. Horne et al., "Enhanced Long-Term Dietary Change and Adherence in a Nutrigenomics-Guided Lifestyle Intervention Compared to a Population-Based (Glb/Dpp) Lifestyle Intervention for Weight Management: Results from the Now Randomised Controlled Trial."
36. Zeevi et al., "Personalized Nutrition by Prediction of Glycemic Responses," *Cell* 163, no. 5 (2015): 1079–94.
37. C. Celis-Morales, J. Lara, and J. C. Mathers, "Personalising Nutritional Guidance for More Effective Behaviour Change," *Proc Nutr Soc* 74, no. 2 (2015): 130-8.

38. C. Celis-Morales et al., "Effect of Personalized Nutrition on Health-Related Behaviour Change: Evidence from the Food4me European Randomized Controlled Trial," *Int J Epidemiol* 46, no. 2 (2017): 578-88.
39. H. L. Tong et al., "Personalized Mobile Technologies for Lifestyle Behavior Change: A Systematic Review, Meta-Analysis, and Meta-Regression," *Prev Med* 148 (2021): 106532.
40. C. Howland and B. Wakefield, "Assessing Telehealth Interventions for Physical Activity and Sedentary Behavior Self-Management in Adults with Type 2 Diabetes Mellitus: An Integrative Review," *Res Nurs Health* 44, no. 1 (2021): 92–110.
41. D. Spruijt-Metz et al., "Innovations in the Use of Interactive Technology to Support Weight Management," *Curr Obes Rep* 4 (2015).
42. L. Wang and C. Miller, "Just-in-the-Moment Adaptive Interventions (Jitai): A Meta-Analytical Review," *Health Commun* 35, no. 12 (2020).
43. W. Hardeman et al., "A Systematic Review of Just-in-Time Adaptive Interventions (Jitais) to Promote Physical Activity," *Int J Behav Nutr Phys Act* 16, no. 1 (2019): 31.
44. O. Perski et al., "Technology-Mediated Just-in-Time Adaptive Interventions (Jitais) to Reduce Harmful Substance Use: A Systematic Review," *Addict* 117, no. 5 (2022).
45. A. Direito et al., "Do Physical Activity and Dietary Smartphone Applications Incorporate Evidence-Based Behaviour Change Techniques?," *BMC Public Health* 14 (2014): 646.
46. N. Aldenaini et al., "Trends in Persuasive Technologies for Physical Activity and Sedentary Behavior: A Systematic Review," *Front Artif Intell* 3 (2020): 7.
47. J. Fanning, S. P. Mullen, and E. McAuley, "Increasing Physical Activity with Mobile Devices: A Meta-Analysis," *J Med Internet Res* 14, no. 6 (2012): e161.
48. I. M. Rosenstock, "The Health Belief Model and Preventive Health Behavior," *Health Educ Monogr* 2, no. 4 (1974): 354–86.
49. I. Ajzen, "The Theory of Planned Behavior," *Organ Behav Hum Decis Process* 50, no. 2 (1991).
50. R. Schwarzer, "Self-Efficacy in the Adoption and Maintenance of Health Behaviors: Theoretical Approaches and a New Model," in *Self-Efficacy: Thought Control of Action.*, 217–43 (Washington, DC: Hemisphere Publishing Corp, 1992).
51. J. D. Fisher and W. A. Fisher, "Changing Aids-Risk Behavior," *Psychol Bull* 111, no. 3 (1992).
52. J.ames O. Prochaska, C. C. DiClemente, and J. C. Norcross, "In Search of How People Change: Applications to Addictive Behaviors," (1997).
53. S. Michie, M. M. van Stralen, and R. West, "The Behaviour Change Wheel: A New Method for Characterising and Designing Behaviour Change Interventions," *Implement Sci* 6 (2011).
54. N. Midjani, F. A. Hossaini, and N. Sharifi, "Promotion of Nutritional Behaviors in the Prevention of Cardiovascular Diseases: Application of the Health Belief Model in Primary Health Care Centers," *BMC Prim Care* 24, no. 1 (2023): 278.
55. M. Saghafi-Asl, S. Aliasgharzadeh, and M. Asghari-Jafarabadi, "Factors Influencing Weight Management Behavior among College Students: An Application of the Health Belief Model," *PLoS One* 15, no. 2 (2020): e0228058.
56. R. Orji, J. Vassileva, and R. Mandryk, "Towards an Effective Health Interventions Design: An Extension of the Health Belief Model," *Online J Public Health Inform* 4, no. 3 (2012): e61050.
57. Orji, Vassileva, and Mandryk, "Towards an Effective Health Interventions Design: An Extension of the Health Belief Model."
58. J. Rise et al., "Predicting the Intention to Quit Smoking and Quitting Behaviour: Extending the Theory of Planned Behaviour," *Br J Health Psychol* 13, no. Pt 2 (2008): 291–310.

59. N. R. Thompson et al., "Theory of Planned Behavior and Perceived Role Model as Predictors of Nutrition and Physical Activity Behaviors among College Students in Health-Related Disciplines," *J Community Health* 45, no. 5 (2020): 965–72.
60. E. J. Kothe and B. A. Mullan, "A Randomised Controlled Trial of a Theory of Planned Behaviour to Increase Fruit and Vegetable Consumption. Fresh Facts," *Appetite* 78 (2014): 68–75.
61. L. Canova, A. Bobbio, and A. M. Manganelli, "Predicting Fruit Consumption: A Multi-Group Application of the Theory of Planned Behavior," *Appetite* 145 (2020): 104490.
62. P. Sheeran, P. M. Gollwitzer, and J. A. Bargh, "Nonconscious Processes and Health," *Health Psychol* 32, no. 5 (2013): 460–73.
63. M. Conner et al., "Some Feelings Are More Important: Cognitive Attitudes, Affective Attitudes, Anticipated Affect, and Blood Donation," *Health Psychol* 32, no. 3 (2013): 264.
64. Schwarzer, "Self-Efficacy in the Adoption and Maintenance of Health Behaviors: Theoretical Approaches and a New Model."
65. J. Parkinson et al., "Health Action Process Approach: Promoting Physical Activity, and Fruit and Vegetable Intake among Australian Adults," *Health Promot Int* 38, no. 4 (2023): daad095.
66. R. J. Williams, T. A. Herzog, and V. N. Simmons, "Risk Perception and Motivation to Quit Smoking: A Partial Test of the Health Action Process Approach," *Addict Behav* 36, no. 7 (2011): 789–91.
67. Fisher and Fisher, "Changing Aids-Risk Behavior."
68. C. Yang et al., "Examining and Adapting the Information-Motivation-Behavioural Skills Model of Medication Adherence among Community-Dwelling Older Patients with Multimorbidity: Protocol for a Cross-Sectional Study," *BMJ Open* 10, no. 3 (2020): e033431.
69. L. A. Nelson et al., "Assessing Barriers to Diabetes Medication Adherence Using the Information-Motivation-Behavioral Skills Model," *Diabetes Res Clin Pract* 142 (2018): 374–84.
70. H.ongbo Jiang et al., "Predictors of Condom Use Behavior among Men Who Have Sex with Men in China Using a Modified Information-Motivation-Behavioral Skills (Imb) Model," *BMC Public Health* 19 (2019): 1–12.
71. S. A John, J. L Walsh, and L. S Weinhardt, "The Information–Motivation–Behavioral Skills Model Revisited: A Network-Perspective Structural Equation Model within a Public Sexually Transmitted Infection Clinic Sample of Hazardous Alcohol Users," *AIDS and Behavior* 21 (2017): 1208–18.
72. J. O. Prochaska and C. C. DiClemente, "Stages and Processes of Self-Change of Smoking: Toward an Integrative Model of Change," *J Consult Clin Psychol* 51, no. 3 (1983): 390–5.
73. J. Adams and M. White, "Why Don't Stage-Based Activity Promotion Interventions Work?," *Health Educ Res* 20, no. 2 (2005): 237–43.
74. Michie, van Stralen, and West, "The Behaviour Change Wheel: A New Method for Characterising and Designing Behaviour Change Interventions."
75. R. Sutherland et al., "A Multicomponent Mhealth-Based Intervention (Swap It) to Decrease the Consumption of Discretionary Foods Packed in School Lunchboxes: Type I Effectiveness-Implementation Hybrid Cluster Randomized Controlled Trial," *J Med Internet Res* 23, no. 6 (2021): e25256.
76. L. K. Chai et al., "Development of Text Messages Targeting Healthy Eating for Children in the Context of Parenting Partnerships," *Nutr Diet* 76, no. 5 (2019): 515–20.
77. C. L. Shoneye et al., "Design and Development of a Digital Weight Management Intervention (Today): Qualitative Study," *JMIR Mhealth Uhealth* 8, no. 9 (2020): e17919.

78. C. Henshall and Z. Davey, "Development of an App for Lung Cancer Survivors (Iexhale) to Increase Exercise Activity and Improve Symptoms of Fatigue, Breathlessness and Depression," *Psychooncology* 29, no. 1 (2020): 139–47.
79. Shoneye et al., "Design and Development of a Digital Weight Management Intervention (Today): Qualitative Study."
80. S. Truelove et al., "The Use of the Behaviour Change Wheel in the Development of Participaction's Physical Activity App," *Prev Med Rep* 20 (2020): 101224.
81. H. Reid et al., "Use of the Behaviour Change Wheel to Improve Everyday Person-Centred Conversations on Physical Activity across Healthcare," *BMC Public Health* 22, no. 1 (2022): 1784.
82. A. Stephenson et al., "Iterative Four-Phase Development of a Theory-Based Digital Behaviour Change Intervention to Reduce Occupational Sedentary Behaviour," *Digit Health* 6 (2020): 2055207620913410.
83. T. S. Wheeler et al., "Feasibility and Usability of an Ontology-Based Mobile Intervention for Patients with Hypertension," *Int J Med Inform* 119 (2018): 8–16.
84. H. M. Jennings et al., "Developing a Theory-Driven Contextually Relevant Mhealth Intervention," *Glob Health Action* 12, no. 1 (2019): 1550736.
85. D. V. Rodriguez et al., "Development of a Computer-Aided Text Message Platform for User Engagement with a Digital Diabetes Prevention Program: A Case Study," *J Am Med Inform Assoc* 29, no. 1 (2021): 155–62.
86. E. A. Fulton et al., "Stopapp: Using the Behaviour Change Wheel to Develop an App to Increase Uptake and Attendance at Nhs Stop Smoking Services," *Healthcare (Basel)* 4, no. 2 (2016): 31.
87. I. Tombor et al., "Development of Smokefree Baby: A Smoking Cessation Smartphone App for Pregnant Smokers," *Transl Behav Med* 6, no. 4 (2016): 533–45.
88. C. Garnett et al., "The Development of Drink Less: An Alcohol Reduction Smartphone App for Excessive Drinkers," *Transl Behav Med* 9, no. 2 (2019): 296–307.
89. M. A. Arden et al., "Development of an Intervention to Increase Adherence to Nebuliser Treatment in Adults with Cystic Fibrosis: Cfhealthhub," *Pilot Feasibility Stud* 7, no. 1 (2021): 1.
90. J. Ribaut et al., "Theory-Driven Development of a Medication Adherence Intervention Delivered by Ehealth and Transplant Team in Allogeneic Stem Cell Transplantation: The Smile Implementation Science Project," *BMC Health Serv Res* 20, no. 1 (2020): 827.
91. H. M. Bowers et al., "Supporting Antidepressant Discontinuation: The Development and Optimisation of a Digital Intervention for Patients in UK Primary Care Using a Theory, Evidence and Person-Based Approach," *BMJ Open* 10, no. 3 (2020): e032312.
92. N O'Brien et al., "Digital Health in Primary Health Care: Current Use and Future Opportunities in the Sub-Saharan African Region," *Imperial College London* (2023).
93. G. Yeoman et al., "Defining Patient Centricity with Patients for Patients and Caregivers: A Collaborative Endeavour," *BMJ Innov* 3, no. 2 (2017): 76-83.
94. J. L. Mair et al., "Effective Behavior Change Techniques in Digital Health Interventions for the Prevention or Management of Noncommunicable Diseases: An Umbrella Review," *Ann Behav Med* 57, no. 10 (2023): 817–35.
95. NICE guideline NG183, "Behaviour Change: Digital and Mobile Health Interventions," National Institute for Health and Care Excellence (2020), https://www.nice.org.uk/guidance/ng183.
96. K. Socha-Dietrich, "Empowering the Health Workforce: Strategies to Make the Most of the Digital Revolution," *OECD Health Working Papers* 129 (2021).
97. B. Furlow, "Information Overload and Unsustainable Workloads in the Era of Electronic Health Records," *Lancet Respir Med* 8, no. 3 (2020): 243–44.

98. G. R. Cohen et al., "Variation in Physicians' Electronic Health Record Documentation and Potential Patient Harm from That Variation," *J Gen Intern Med* 34, no. 11 (2019): 2355–67.
99. "What Is Telehealth?," *Catalyst Carryover* 4, no. 1 (2018).
100. T. A.Lustig, *The Role of Telehealth in an Evolving Health Care Environment: Workshop Summary* (Washington, DC: National Academies Press, 2012).
101. A. Blandford et al., "Opportunities and Challenges for Telehealth within, and Beyond, a Pandemic," *Lancet Glob Health* 8, no. 11 (2020): e1364–e65.
102. World Health Organization, "Mhealth: Use of Mobile Wireless Technologies for Public Health (Eb139/8)," in *Provisional Agenda Item* (Geneva: World Health Organisation, 2016, https://apps.who.int/gb/ebwha/pdf_files/EB139/B139_8-en.pdf).
103. S. A. Buckingham et al., "Mobile Health Interventions to Promote Physical Activity and Reduce Sedentary Behaviour in the Workplace: A Systematic Review," *Digit Health* 5 (2019): 2055207619839883.
104. V. J. Block et al., "Continuous Daily Assessment of Multiple Sclerosis Disability Using Remote Step Count Monitoring," *J Neurol* 264, no. 2 (2017): 316–26.
105. C. Friedman, "Remote Monitoring Support Services for People with Intellectual and Developmental Disabilities," *J Policy Pract Intellect Disabil* 20, no. 3 (2023): 298–307.
106. M. J. Tasse, J. B. Wagner, and M. Kim, "Using Technology and Remote Support Services to Promote Independent Living of Adults with Intellectual Disability and Related Developmental Disabilities," *J Appl Res Intellect Disabil* 33, no. 3 (2020): 640–47.
107. Y. Gu et al., "Predicting Medication Adherence Using Ensemble Learning and Deep Learning Models with Large Scale Healthcare Data," *Sci Rep* 11, no. 1 (2021): 18961.
108. B. Singh et al., "Systematic Review and Meta-Analysis of the Effectiveness of Chatbots on Lifestyle Behaviours," *NPJ Digit Med* 6, no. 1 (2023): 118.
109. L. V. Eshuis et al., "Efficacy of Immersive PTSD Treatments: A Systematic Review of Virtual and Augmented Reality Exposure Therapy and a Meta-Analysis of Virtual Reality Exposure Therapy," *J Psychiatr Res* 143 (2021): 516–27.
110. I. Phelan et al., "Immersive Virtual Reality in Children with Upper Limb Injuries: Findings from a Feasibility Study," *J Pediatr Rehabil Med* 14, no. 3 (2021): 401–14.
111. B. Serrano et al., "Virtual Reality and Anxiety Disorders Treatment: Evolution and Future Perspectives," in *Virtual Reality for Psychological and Neurocognitive Interventions*, ed. Albert "Skip" Rizzo and Stéphane Bouchard, Virtual Reality Technologies for Health and Clinical Applications, 47-84. (New York, NY: Springer New York, 2019).
112. D. Wakabayashi, "Google and the University of Chicago Are Sued over Data Sharing," The New York Times (2019), https://www.nytimes.com/2019/06/26/technology/google-university-chicago-data-sharing-lawsuit.html (Accessed 13/06/2024).
113. D. Muoio and A. Burky, "Advocate Aurora, Wakemed Get Served with Class Action over Meta's Alleged Patient Data Mining," Fierce Healthcare (2022), https://www.fiercehealthcare.com/health-tech/report-third-top-hospitals-websites-collecting-patient-data-facebook.
114. Ari. B. Friedman et al., "Widespread Third-Party Tracking on Hospital Websites Poses Privacy Risks for Patients and Legal Liability for Hospitals: Study Examines the Widespread Use of Third-Party Tracking on Hospital Websites and the Privacy Risks for Patients and Legal Liability for Hospitals," *Health Aff* 42, no. 4 (2023): 508–15.
115. R. Thomas, "NHS Cyber Attack Causing Chaos in London Hits Other Hospitals in UK," The Independent (2024), https://www.independent.co.uk/news/health/nhs-russian-cyber-attack-london-hospital-portsmouth-b2559893.html (Accessed 13/06/2024).

116. World Health Organization, "Mhealth: Use of Mobile Wireless Technologies for Public Health (Eb139/8)."
117. J. Shaw et al., "Digital Health Technologies for More Equitable Health Systems: A Discussion Paper," Toronto (ON): University of Toronto Joint Centre for Bioethics, Women's College Hospital Institute for Health Systems Solutions and Virtual Care (2020), https://www.womensacademics.ca/wp-content/uploads/2022/11/Digital_Health_and_Equity_Report__September_2020FINAL.pdf (Accessed 30/09/2024).
118. S. P. Rowland et al., "Digital Health Technology-Specific Risks for Medical Malpractice Liability," *NPJ Digit Med* 5, no. 1 (2022): 157.
119. Rowland et al., "Digital Health Technology-Specific Risks for Medical Malpractice Liability."
120. G. Tangari et al., "Mobile Health and Privacy: Cross Sectional Study," *BMJ* 373 (2021): n1248.
121. J. P. M. Faessen et al., "Eating for 2: A Systematic Review of Dutch App Stores for Apps Promoting a Healthy Diet During Pregnancy," *Curr Dev Nutr* 6, no. 6 (2022): nzac087.
122. J. Melcher and J. Torous, "Smartphone Apps for College Mental Health: A Concern for Privacy and Quality of Current Offerings," *Psychiatr Serv* 71, no. 11 (2020): 1114–19.
123. S. Grock et al., "A Review of Technology-Assisted Interventions for Diabetes Prevention," *Curr Diab Rep* 17, no. 11 (2017): 107.
124. A. Stephenson et al., "Using Computer, Mobile and Wearable Technology Enhanced Interventions to Reduce Sedentary Behaviour: A Systematic Review and Meta-Analysis," *Int J Behav Nutr Phys Act* 14, no. 1 (2017): 105.
125. M. H. Jarrahi, N. Gafinowitz, and G. Shin, "Activity Trackers, Prior Motivation, and Perceived Informational and Motivational Affordances," *Pers Ubiquitous Comput* 22, no. 2 (2017): 433–48.
126. R. A. J. de Vries et al., "A Word of Advice: How to Tailor Motivational Text Messages Based on Behavior Change Theory to Personality and Gender," *Pers Ubiquitous Comput* 21, no. 4 (2017): 675–87.
127. Stephenson et al., "Using Computer, Mobile and Wearable Technology Enhanced Interventions to Reduce Sedentary Behaviour: A Systematic Review and Meta-Analysis."
128. W. Hardeman et al., "A Systematic Review of Just-in-Time Adaptive Interventions (Jitais) to Promote Physical Activity," *Int J Behav Nutr Phys Act*16 (2019).
129. F.. Naughton, "Delivering 'Just-in-Time' Smoking Cessation Support Via Mobile Phones: Current Knowledge and Future Directions," *Nicotine Tob Res* 19, no. 3 (2017).
130. Hardeman et al., "A Systematic Review of Just-in-Time Adaptive Interventions (Jitais) to Promote Physical Activity."
131. D. H. Gustafson et al., "A Smartphone Application to Support Recovery from Alcoholism: A Randomized Clinical Trial," *JAMA Psychiatry* 71, no. 5 (2014): 566–72.
132. M. J. Yang et al., "A Just-in-Time Adaptive Intervention (Jitai) for Smoking Cessation: Feasibility and Acceptability Findings," *Addict Behav* 136 (2023): 107467.
133. R. Sutton and A. Barto, *Reinforcement Learning: An Introduction*, Second ed. (Cambridge, MA: The MIT Press, 2018).
134. P. Paredes et al., "Poptherapy: Coping with Stress through Pop-Culture," *Proceedings of the 8th International Conference on Pervasive Computing Technologies for Healthcare* (2014).
135. A. Guez et al., "Adaptive Treatment of Epilepsy Via Batch-Mode Reinforcement Learning" (Paper Presented at the AAAI, 2008).
136. J. C. Lauffenburger et al., "Reinforcement Learning to Improve Non-Adherence for Diabetes Treatments by Optimising Response and Customising Engagement (Reinforce): Study Protocol of a Pragmatic Randomised Trial," *BMJ Open* 11, no. 12 (2021): e052091.

137. A. Aswani et al., "Behavioral Modeling in Weight Loss Interventions," *Eur J Oper Res* 272, no. 3 (2019): 1058–72.
138. D. Ganguly et al., "Outcome Prediction from Behaviour Change Intervention Evaluations Using a Combination of Node and Word Embedding," *AMIA Annu Symp Proc* 2021 (2021): 486.
139. T. W. Bickmore et al., "A Randomized Controlled Trial of an Automated Exercise Coach for Older Adults," *J Am Geriatr Soc* 61, no. 10 (2013): 1676–83.
140. N. Stein and K. Brooks, "A Fully Automated Conversational Artificial Intelligence for Weight Loss: Longitudinal Observational Study among Overweight and Obese Adults," *JMIR Diabetes* 2, no. 2 (2017): e28.
141. P.ul Resnick and H. R. Varian, "Recommender Systems," *Communications of the ACM* 40, no. 3 (1997): 56–58.
142. G. A. Sielis, A. Tzanavari, and G. A. Papadopoulos, "Recommender Systems Review of Types, Techniques, and Applications," in *Encyclopedia of Information Science and Technology, Third Edition* (IGI global, 2015).
143. J. O'Donovan and B. Smyth, "Trust in Recommender Systems," in *Proceedings of the 10th International Conference on Intelligent User Interfaces* (San Diego, CA: Association for Computing Machinery, 2005).
144. G. Adomavicius and A. Tuzhilin, "Context-Aware Recommender Systems," in *Recommender Systems Handbook*, 217–53. (Springer, 2010).
145. F. Gräßer et al., "A Pharmaceutical Therapy Recommender System Enabling Shared Decision-Making," *User Modeling and User-Adapted Interaction* 32, no. 5 (2021): 1019–62.
146. L. Verboven et al., "A Treatment Recommender Clinical Decision Support System for Personalized Medicine: Method Development and Proof-of-Concept for Drug Resistant Tuberculosis," *BMC Med Inform Decis Mak* 22, no. 1 (2022): 56.
147. S. H. Oh et al., "Reinforcement Learning-Based Expanded Personalized Diabetes Treatment Recommendation Using South Korean Electronic Health Records," *Expert Syst Appl* 206 (2022): 117932.
148. M. Waqar et al., "An Adaptive Doctor-Recommender System," *Behav Inf Technol* 38, no. 9 (2019): 959–73.
149. B. Rahdari et al., "Helper: An Interactive Recommender System for Ovarian Cancer Patients and Caregivers" (Paper presented at the Proceedings of the 16th ACM Conference on Recommender Systems, 2022).
150. J. Ni, L. Muhlstein, and J. McAuley, "Modeling Heart Rate and Activity Data for Personalized Fitness Recommendation" (Paper presented at the WWW 2019: The World Wide Web Conference, New York, 2019).
151. A. G. Mahyari and P. Pirolli, "Physical Exercise Recommendation and Success Prediction Using Interconnected Recurrent Neural Networks," *arXiv* (2020): 1–9.
152. Z. Zhao et al., "Effects of a Personalized Fitness Recommender System Using Gamification and Continuous Player Modeling: System Design and Long-Term Validation Study," *JMIR Serious Games* 8, no. 4 (2020): e19968.
153. M. Rabbi et al., "Automated Personalized Feedback for Physical Activity and Dietary Behavior Change with Mobile Phones: A Randomized Controlled Trial on Adults," *JMIR Mhealth Uhealth* 3, no. 2 (2015): e42.
154. G. M. Tanimul Ahsan et al., "Toward an Mhealth Intervention for Smoking Cessation," *2013 IEEE 37th Annual Computer Software and Applications Conference Workshops* (2013): 345–50.
155. M. Khwaja et al., "Aligning Daily Activities with Personality: Towards a Recommender System for Improving Wellbeing" (Paper Presented at the Proceedings of the 13th ACM Conference on Recommender Systems, 2019).

156. C. H. Chen et al., "Person-Personalized Expert Recommendation System for Optimized Nutrition," *IEEE Trans Biomed Circuits Syst* 12, no. 1 (2018): 151–60.
157. A. Vaswani et al., "Attention Is All You Need," *Advances in neural information processing systems* 30 (2017).
158. S. Lim and R. Schmälzle, "Artificial Intelligence for Health Message Generation: An Empirical Study Using a Large Language Model (LLM) and Prompt Engineering," *Front commun* 8 (2023): 1129082.
159. Z. Kraljevic et al., "Foresight – A Generative Pretrained Transformer for Modelling of Patient Timelines Using Electronic Health Records: A Retrospective Modelling Study," *Lancet Digit Health* 6, no. 4 (2024).
160. A. Sarraju et al., "Appropriateness of Cardiovascular Disease Prevention Recommendations Obtained from a Popular Online Chat-Based Artificial Intelligence Model," *JAMA* 329, no. 10 (2023): 842–44.
161. Y. Matias, "Our Progress on Generative Ai in Healthcare," Google (2024), https://blog.google/technology/health/google-generative-ai-healthcare/ (Accessed 16/06/2024).
162. A. Davies and J. Mueller, "Designing an Mhealth Intervention," in *Developing Medical Apps and Mhealth Interventions*, ed. Alan Davies and Julia Mueller, Health Informatics, 75–110. (Cham: Springer International Publishing, 2020).
163. K. Skivington et al., "A New Framework for Developing and Evaluating Complex Interventions: Update of Medical Research Council Guidance," *BMJ* 374 (2021): n2061.
164. Davies and Mueller, "Designing an Mhealth Intervention."
165. J. Cane, D. O'Connor, and S. Michie, "Validation of the Theoretical Domains Framework for Use in Behaviour Change and Implementation Research," *Implement Sci* 7, no. 37 (2012): 37.
166. Michie, van Stralen, and West, "The Behaviour Change Wheel: A New Method for Characterising and Designing Behaviour Change Interventions."
167. S. Michie and L. Atkins, *The Behaviour Change Wheel: A Guide to Designing Interventions* (London: Silverback Publishing, 2014).
168. S. Michie et al., "The Behavior Change Technique Taxonomy (v1) of 93 Hierarchically Clustered Techniques: Building an International Consensus for the Reporting of Behavior Change Interventions," *Ann Behav Med* 46, no. 1 (2013): 81–95.
169. M. M. Marques et al., "The Behaviour Change Technique Ontology: Transforming the Behaviour Change Technique Taxonomy v1," *Wellcome Open Res* 8 (2023): 308.
170. Marques et al., "The Behaviour Change Technique Ontology: Transforming the Behaviour Change Technique Taxonomy v1."
171. I.l Nahum-Shani et al., "Just-in-Time Adaptive Interventions (Jitais) in Mobile Health: Key Components and Design Principles for Ongoing Health Behavior Support," *Annals of Behavioral Medicine* (2018): 1–17.
172. L. M. Collins, S. A. Murphy, and K. L. Bierman, "A Conceptual Framework for Adaptive Preventive Interventions," *Prev Sci* 5, no. 3 (2004): 185–96.
173. Nahum-Shani et al., "Just-in-Time Adaptive Interventions (Jitais) in Mobile Health: Key Components and Design Principles for Ongoing Health Behavior Support."
174. L. Yardley et al., "The Person-Based Approach to Intervention Development: Application to Digital Health-Related Behavior Change Interventions," *J Med Internet Res* 17, no. 1 (2015): e30.
175. Davies and Mueller, "Designing an Mhealth Intervention."
176. S. Michie et al., "Developing and Evaluating Digital Interventions to Promote Behavior Change in Health and Health Care: Recommendations Resulting from an International Workshop," *J Med Internet Res* 19, no. 6 (2017): e232.
177. J. Farao et al., "A User-Centred Design Framework for Mhealth," *PloS One* 15, no. 8 (2020): e0237910.

178. J. E. van Gemert-Pijnen et al., "A Holistic Framework to Improve the Uptake and Impact of Ehealth Technologies," *J Med Internet Res* 13, no. 4 (2011): e111.
179. S. Michie et al., "Effective Techniques in Healthy Eating and Physical Activity Interventions: A Meta-Regression," *Health Psych* 28, no. 6 (2009): 690.
180. H. Schroé et al., "Which Behaviour Change Techniques Are Effective to Promote Physical Activity and Reduce Sedentary Behaviour in Adults: A Factorial Randomized Trial of an E-and M-Health Intervention," *Int J Behav Nutr Phys Act* 17 (2020): 1–16.
181. S. Compernolle et al., "Effectiveness of Interventions Using Self-Monitoring to Reduce Sedentary Behavior in Adults: A Systematic Review and Meta-Analysis," *Int J Behav Nutr Phys Act* 16, no. 1 (2019): 63.
182. E. J. Lyons and M. C. Swartz, "Motivational Dynamics of Wearable Activity Monitors," *ACSM's Health Fit J* 21, no. 5 (2017): 21–26.
183. D. Kwasnicka et al., "Theoretical Explanations for Maintenance of Behaviour Change: A Systematic Review of Behaviour Theories," *Health Psychol Rev* 10, no. 3 (2016): 277–96.
184. L. Xu et al., "The Effects of Mhealth-Based Gamification Interventions on Participation in Physical Activity: Systematic Review," *JMIR Mhealth Uhealth* 10, no. 2 (2022): e27794.
185. E. A. Edwards et al., "Gamification for Health Promotion: Systematic Review of Behaviour Change Techniques in Smartphone Apps," *BMJ Open* 6, no. 10 (2016): e012447.
186. M. S. Patel et al., "Effect of Goal-Setting Approaches within a Gamification Intervention to Increase Physical Activity among Economically Disadvantaged Adults at Elevated Risk for Major Adverse Cardiovascular Events: The Engage Randomized Clinical Trial," *JAMA Cardiol* 6, no. 12 (2021): 1387–96.
187. M. S. Patel et al., "Effectiveness of Behaviorally Designed Gamification Interventions with Social Incentives for Increasing Physical Activity among Overweight and Obese Adults across the United States: The Step up Randomized Clinical Trial," *JAMA Intern Med* 179, no. 12 (2019): 1624–32.
188. M. Aitken and D. Nass, "Digital Health Trends 2021: Innovation, Evidence, Regulation, and Adoption," The IQVIA Institute (2021), https://www.iqvia.com/insights/the-iqvia-institute/reports-and-publications/reports/digital-health-trends-2021 (Accessed 30/09/2024).
189. Edwards et al., "Gamification for Health Promotion: Systematic Review of Behaviour Change Techniques in Smartphone Apps."
190. Skivington et al., "A New Framework for Developing and Evaluating Complex Interventions: Update of Medical Research Council Guidance."; Aitken and Nass, "Digital Health Trends 2021: Innovation, Evidence, Regulation, and Adoption."
191. P. Voorheis et al., "Integrating Behavioral Science and Design Thinking to Develop Mobile Health Interventions: Systematic Scoping Review," *JMIR Mhealth Uhealth* 10, no. 3 (2022): e35799.
192. Y. Wang et al., "A Systematic Review of Application and Effectiveness of Mhealth Interventions for Obesity and Diabetes Treatment and Self-Management," *Adv Nutr* 8, no. 3 (2017): 449–62.
193. S. Zheng et al., "Effectiveness of Holistic Mobile Health Interventions on Diet, and Physical, and Mental Health Outcomes: A Systematic Review and Meta-Analysis," *EClinicalMedicine* 66 (2023): 102309.
194. Xu et al., "The Effects of Mhealth-Based Gamification Interventions on Participation in Physical Activity: Systematic Review."
195. Wang et al., "A Systematic Review of Application and Effectiveness of Mhealth Interventions for Obesity and Diabetes Treatment and Self-Management."

196. Zheng et al., "Effectiveness of Holistic Mobile Health Interventions on Diet, and Physical, and Mental Health Outcomes: A Systematic Review and Meta-Analysis."
197. Wang et al., "A Systematic Review of Application and Effectiveness of Mhealth Interventions for Obesity and Diabetes Treatment and Self-Management."
198. Y. J. Oh et al., "A Systematic Review of Artificial Intelligence Chatbots for Promoting Physical Activity, Healthy Diet, and Weight Loss," *Int J Behav Nutr Phys Act* 18, no. 1 (2021): 160.
199. Singh et al., "Systematic Review and Meta-Analysis of the Effectiveness of Chatbots on Lifestyle Behaviours."
200. Kwasnicka et al., "Theoretical Explanations for Maintenance of Behaviour Change: A Systematic Review of Behaviour Theories."
201. D. Wang et al., "Association between Socioeconomic Status and Health Behaviour Change before and after Non-Communicable Disease Diagnoses: A Multicohort Study," *Lancet Public Health* 7, no. 8 (2022): e670–e82.
202. Davies and Mueller, "Designing an Mhealth Intervention."
203. J. P. Ioannidis, "Effect of the Statistical Significance of Results on the Time to Completion and Publication of Randomized Efficacy Trials," *JAMA* 279, no. 4 (1998): 281–6.
204. C. McCallum, J. Rooksby, and C. M. Gray, "Evaluating the Impact of Physical Activity Apps and Wearables: Interdisciplinary Review," *JMIR Mhealth Uhealth* 6, no. 3 (2018): e58.
205. C. Guo et al., "Challenges for the Evaluation of Digital Health Solutions – A Call for Innovative Evidence Generation Approaches," *NPJ Digit Med* 3, no. 1 (2020): 110.
206. Guo et al., "Challenges for the Evaluation of Digital Health Solutions – A Call for Innovative Evidence Generation Approaches."
207. A. A. Montgomery, T. J. Peters, and P. Little, "Design, Analysis and Presentation of Factorial Randomised Controlled Trials," *BMC Med Res Methodol* 3, no. 1 (2003): 26.
208. H. Schroe et al., "A Factorial Randomised Controlled Trial to Identify Efficacious Self-Regulation Techniques in an E- and M-Health Intervention to Target an Active Lifestyle: Study Protocol," *Trials* 20, no. 1 (2019): 340.
209. M. Arshed et al., "Efficacy of Mhealth and Education-Led Peer Counseling for Patients with Hypertension and Coronary Artery Disease in Pakistan: Study Protocol for a Double-Blinded Pragmatic Randomized-Controlled Trial with Factorial Design," *Trials* 24 (2023): 448.
210. B. Spring et al., "A Factorial Experiment to Optimize Remotely Delivered Behavioral Treatment for Obesity: Results of the Opt-in Study," *Obesity (Silver Spring)* 28, no. 9 (2020): 1652–62.
211. P. Klasnja et al., "Microrandomized Trials: An Experimental Design for Developing Just-in-Time Adaptive Interventions," *Health Psychol* 34S, no. 0 (2015): 1220–8.
212. P. Liao et al., "Sample Size Calculations for Micro-Randomized Trials in Mhealth," *Stat Med* 35, no. 12 (2016); Klasnja et al., "Microrandomized Trials: An Experimental Design for Developing Just-in-Time Adaptive Interventions."
213. P. Klasnja et al., "Efficacy of Contextually Tailored Suggestions for Physical Activity: A Micro-Randomized Optimization Trial of Heartsteps," *Ann Behav Med* 53, no. 6 (2019): 573–82.
214. A. Aguilera et al., "Mhealth App Using Machine Learning to Increase Physical Activity in Diabetes and Depression: Clinical Trial Protocol for the Diamante Study," *BMJ Open* 10, no. 8 (2020): e034723.
215. C. A. Figueroa et al., "A Text Messaging Intervention for Coping with Social Distancing During Covid-19 (Staywell at Home): Protocol for a Randomized Controlled Trial," *JMIR Res Protoc* 10, no. 1 (2021): e23592.

216. J. Wang et al., "Effectiveness of Gamified Team Competition as Mhealth Intervention for Medical Interns: A Cluster Micro-Randomized Trial," *NPJ Digit Med* 6, no. 1 (2023): 4.
217. M. Rabbi et al., "Toward Increasing Engagement in Substance Use Data Collection: Development of the Substance Abuse Research Assistant App and Protocol for a Microrandomized Trial Using Adolescents and Emerging Adults," *JMIR Res Protoc* 7, no. 7 (2018): e166.
218. X. Liu, N. Deliu, and B. Chakraborty, "Microrandomized Trials: Developing Just-in-Time Adaptive Interventions for Better Public Health," *Am J Public Health* 113, no. 1 (2023): 60–69.
219. L. M. Collins, S. A. Murphy, and V. Strecher, "The Multiphase Optimization Strategy (MOST) and the Sequential Multiple Assignment Randomized Trial (SMART): New Methods for More Potent eHealth Interventions," *Am J Prev Med* 32, no. 5 Suppl (2007): S112–8.
220. B. B. Gonze et al., "Use of a Smartphone App to Increase Physical Activity Levels in Insufficiently Active Adults: Feasibility Sequential Multiple Assignment Randomized Trial (Smart)," *JMIR Res Protoc* 9, no. 10 (2020): e14322.
221. E. K. Czyz et al., "Adaptive Intervention for Prevention of Adolescent Suicidal Behavior after Hospitalization: A Pilot Sequential Multiple Assignment Randomized Trial," *J Child Psychol Psychiatry* 62, no. 8 (2021): 1019–31.
222. S. D. Lambert et al., "Adaptive Web-Based Stress Management Programs among Adults with a Cardiovascular Disease: A Pilot Sequential Multiple Assignment Randomized Trial (Smart)," *Patient Educ Couns* 105, no. 6 (2022): 1587–97.
223. X. Yan et al., "Sequential Multiple Assignment Randomized Trial (Smart) to Identify Optimal Sequences of Telemedicine Interventions for Improving Initiation of Insulin Therapy: A Simulation Study," *BMC Med Res Methodol* 21, no. 1 (2021): 200.
224. Guo et al., "Challenges for the Evaluation of Digital Health Solutions – A Call for Innovative Evidence Generation Approaches."
225. M. J. Davies et al., "Management of Hyperglycemia in Type 2 Diabetes, 2018. A Consensus Report by the American Diabetes Association (Ada) and the European Association for the Study of Diabetes (Easd)," *Diabetes Care* 41, no. 12 (2018): 2669–701.
226. C. Howland and B. Wakefield, "Assessing Telehealth Interventions for Physical Activity and Sedentary Behavior Self-Management in Adults with Type 2 Diabetes Mellitus: An Integrative Review," *Res Nurs Health* 44, no. 1 (2021).
227. H. L. Tong et al., "Personalized Mobile Technologies for Lifestyle Behavior Change: A Systematic Review, Meta-Analysis, and Meta-Regression," *Prev Med* 148 (2021).
228. Hardeman et al., "A Systematic Review of Just-in-Time Adaptive Interventions (Jitais) to Promote Physical Activity."
229. Faessen et al., "Eating for 2: A Systematic Review of Dutch App Stores for Apps Promoting a Healthy Diet During Pregnancy."
230. F. Monteiro-Guerra et al., "Personalization in Real-Time Physical Activity Coaching Using Mobile Applications: A Scoping Review," *IEEE J Biomed Health Inform* 24, no. 6 (2020): 1738-51.
231. Tong et al., "Personalized Mobile Technologies for Lifestyle Behavior Change: A Systematic Review, Meta-Analysis, and Meta-Regression."
232. Tong et al., "Personalized Mobile Technologies for Lifestyle Behavior Change: A Systematic Review, Meta-Analysis, and Meta-Regression."
233. A. L. Orsama et al., "Active Assistance Technology Reduces Glycosylated Hemoglobin and Weight in Individuals with Type 2 Diabetes: Results of a Theory-Based Randomized Trial," *Diabetes Technol Ther* 15, no. 8 (2013): 662-9.

234. E. Yom-Tov et al., "Encouraging Physical Activity in Patients with Diabetes: Intervention Using a Reinforcement Learning System," *J Med Internet Res* 19, no. 10 (2017).
235. Monteiro-Guerra et al., "Personalization in Real-Time Physical Activity Coaching Using Mobile Applications: A Scoping Review."
236. Direito et al., "Do Physical Activity and Dietary Smartphone Applications Incorporate Evidence-Based Behaviour Change Techniques?"
237. Hardeman et al., "A Systematic Review of Just-in-Time Adaptive Interventions (Jitais) to Promote Physical Activity."
238. Tong et al., "Personalized Mobile Technologies for Lifestyle Behavior Change: A Systematic Review, Meta-Analysis, and Meta-Regression."
239. Monteiro-Guerra et al., "Personalization in Real-Time Physical Activity Coaching Using Mobile Applications: A Scoping Review."
240. Monteiro-Guerra et al., "Personalization in Real-Time Physical Activity Coaching Using Mobile Applications: A Scoping Review."
241. Michie et al., "The Behavior Change Technique Taxonomy (V1) of 93 Hierarchically Clustered Techniques: Building an International Consensus for the Reporting of Behavior Change Interventions."
242. J. M. Friedman, "Leptin and the Endocrine Control of Energy Balance," *Nat Metab* 1, no. 8 (2019): 754–64.
243. T. A. Wadden et al., "Short- and Long-Term Changes in Serum Leptin Dieting Obese Women: Effects of Caloric Restriction and Weight Loss," *J Clin Endocrinol Metab* 83, no. 1 (1998): 214–8.
244. C. Sudlow et al., "UK Biobank: An Open Access Resource for Identifying the Causes of a Wide Range of Complex Diseases of Middle and Old Age," *PLoS Med* 12, no. 3 (2015): e1001779.
245. L Avery et al., "Barriers and Enabling Factors to Use of a Physical Activity Behavioural Intervention for Adults with Type 2 Diabetes Delivered in Primary Care: Qualitative Findings from the Movement as Medicine Open Pilot Study" (paper presented at the Diabetes UK Professional Conference, 2015).
246. D. Casey, M. De Civita, and K. Dasgupta, "Understanding Physical Activity Facilitators and Barriers During and Following a Supervised Exercise Programme in Type 2 Diabetes: A Qualitative Study," *Diabet Med* 27, no. 1 (2010): 79–84.
247. K. A. Cradock et al., "Identifying Barriers and Facilitators to Diet and Physical Activity Behaviour Change in Type 2 Diabetes Using a Design Probe Methodology," *J Pers Med* 11, no. 2 (2021): 72.
248. D. Erickson, "Barriers to Physical Activity in People with Type 2 Diabetes Enrolled in a Worksite Diabetes Disease Management Program," *Diabetes Educ* 39, no. 5 (2013): 626-34.
249. C. M. Swoboda, C. K. Miller, and C. E. Wills, "Frequency of Diet and Physical Activity Goal Attainment and Barriers Encountered among Adults with Type 2 Diabetes During a Telephone Coaching Intervention," *Clin Diabetes* 35, no. 5 (2017): 286-93.
250. Casey, De Civita, and Dasgupta, "Understanding Physical Activity Facilitators and Barriers During and Following a Supervised Exercise Programme in Type 2 Diabetes: A Qualitative Study."
251. Cane, O'Connor, and Michie, "Validation of the Theoretical Domains Framework for Use in Behaviour Change and Implementation Research."
252. People with T2D in the UK Biobank were previously stratified into clusters, and the reader is referred to Chapter 2 for more details.

253. S. Raschka, "Mlxtend: Providing Machine Learning and Data Science Utilities and Extensions to Python's Scientific Computing Stack," *J Open Source Softw* 3, no. 24 (2018): 638.
254. F. Pedregosa et al., "Scikit-Learn: Machine Learning in Python," *J Mach Learn Res* 12, no. 85 (2011): 2825–30.
255. M. Mars et al., "Fasting Leptin and Appetite Responses Induced by a 4-Day 65%-Energy-Restricted Diet," *Int J Obes (Lond)* 30, no. 1 (2006): 122–8.
256. N. Katsiki, D. P. Mikhailidis, and M. Banach, "Leptin, Cardiovascular Diseases and Type 2 Diabetes Mellitus," *Acta Pharmacol Sin* 39, no. 7 (2018): 1176–88.
257. A. Doherty et al., "Gwas Identifies 14 Loci for Device-Measured Physical Activity and Sleep Duration," *Nat Commun* 9, no. 1 (2018): 5257.
258. M. Willetts et al., "Statistical Machine Learning of Sleep and Physical Activity Phenotypes from Sensor Data in 96,220 UK Biobank Participants," *Sci Rep* 8, no. 1 (2018): 7961.
259. S. F. Chastin et al., "Combined Effects of Time Spent in Physical Activity, Sedentary Behaviors and Sleep on Obesity and Cardio-Metabolic Health Markers: A Novel Compositional Data Analysis Approach," *PLoS One* 10, no. 10 (2015): e0139984.
260. D. Dumuid et al., "Compositional Data Analysis for Physical Activity, Sedentary Time and Sleep Research," *Stat Methods Med Res* 27, no. 12 (2018): 3726–38.
261. Dumuid et al., "Compositional Data Analysis for Physical Activity, Sedentary Time and Sleep Research."
262. J. A. Damen et al., "Prediction Models for Cardiovascular Disease Risk in the General Population: Systematic Review," *BMJ* 353 (2016): i2416.
263. C. Davidson-Pilon, "Lifelines: Survival Analysis in Python," *J Open Source Softw* 4, no. 40 (2019): 1317.
264. E. A. Finkelstein et al., "Effectiveness of Activity Trackers with and without Incentives to Increase Physical Activity (Trippa): A Randomised Controlled Trial," *Lancet Diabetes Endocrinol* 4, no. 12 (2016): 983–95.
265. J. Meyer et al., "Identification and Classification of Usage Patterns in Long-Term Activity Tracking" (Paper Presented at the Proceedings of the 2017 CHI Conference on Human Factors in Computing Systems, New York, 2017).
266. H. Zuo et al., "Association between Serum Leptin Concentrations and Insulin Resistance: A Population-Based Study from China," *PLoS One* 8, no. 1 (2013): e54665.
267. M. V. Fedewa et al., "The Effect of Chronic Exercise Training on Leptin: A Systematic Review and Meta-Analysis of Randomized Controlled Trials," *Sports Med* 48, no. 6 (2018): 1437–50.
268. Finkelstein et al., "Effectiveness of Activity Trackers with and without Incentives to Increase Physical Activity (Trippa): A Randomised Controlled Trial."
269. Meyer et al., "Identification and Classification of Usage Patterns in Long-Term Activity Tracking."
270. Finkelstein et al., "Effectiveness of Activity Trackers with and without Incentives to Increase Physical Activity (Trippa): A Randomised Controlled Trial."
271. Meyer et al., "Identification and Classification of Usage Patterns in Long-Term Activity Tracking."
272. G. J. Hollands et al., "The Impact of Communicating Genetic Risks of Disease on Risk-Reducing Health Behaviour."
273. A. K. Akobeng, "Understanding Randomised Controlled Trials," *Arch Dis Child* 90, no. 8 (2005): 840–4.
274. F. R. Cavallo et al., "A Point-of-Care Device for Fully Automated, Fast and Sensitive Protein Quantification Via QPCR," *Biosensors (Basel)* 12, no. 7 (2022): 537.

275. F. R. Cavallo et al., "Aptasensor for Quantification of Leptin through PCR Amplification of Short DNA-Aptamers," *ACS Sens* 6, no. 3 (2021): 709-715.
276. K. Balog and T. Kenter, "Personal Knowledge Graphs: A Research Agenda" (paper presented at the 2019 ACM SIGIR International Conference on Theory of Information Retrieval, New York, 2019).
277. X. Guo, X. Zhang, and Y. Sun, "The Privacy–Personalization Paradox in Mhealth Services Acceptance of Different Age Groups," *Electron Commer Res Appl* 16 (2016): 55–65.
278. Tangari et al., "Mobile Health and Privacy: Cross Sectional Study."

4 Biosensors and point-of-care devices

BIOMARKERS AND PERSONALISED MEDICINE

Biomarkers are defined as objective measures of biological, physiological, and pathogenic processes or response to a therapeutic intervention.[1] Biomarkers can be categorised into three broad types: molecular, such as chemicals, proteins, or genes; cellular, including cell type, cell morphology, and tissue histology; or imaging, such as x-ray, computed tomography (CT), positron emission tomography (PET), or magnetic resonance imaging (MRI) features. Biomarkers can also be derived through other sensing technologies such as electrocardiograms and encephalograms. Recently, thanks to advances in digital and wearable technologies and the rapid increase in data collected from individuals, a fourth category of biomarker has appeared: digital biomarkers. Digital biomarkers are possible as a result of devices such as wearable sensors and smartphones that can collect objective measures of physical and cognitive function. For example, data collected from the interaction between patients and smartphones can serve as biomarkers for several neuropsychological constructs such as memory, executive function, language, and intelligence, which can all measure a patient's cognitive function.[2]

Biomarkers can also be classified based on their application. In 2016, the US Food and Drug Administration, together with the National Institutes of Health, published a classification of biomarkers according to six categories: diagnostic, monitoring, pharmacodynamic, predictive, prognostic, and safety biomarkers.[3]

Diagnostic biomarkers are used to detect or confirm the presence of a disease, and to identify individuals with a subtype of the disease. Common examples of diagnostic biomarkers include HbA1c to diagnose type 2 diabetes mellitus,[4] and blood pressure to diagnose essential hypertension.[5]

Monitoring biomarkers are measured periodically to assess the status of a disease and its progression, as in the example of cancer antigen 125, which can be used to assess disease status during and after treatment for ovarian cancer.[6] Detecting and quantifying diagnostic and monitoring biomarkers can improve health outcomes by providing precise and timely insights into a patient's health status.[7] Additionally, monitoring biomarkers can offer insights into disease progression, enabling proactive interventions that can slow or prevent complications associated with conditions.

Pharmacodynamic/response biomarkers change in response to a medical substance or an environmental agent. Blood pressure can be a pharmacodynamic biomarker if used to assess the response to pressure-lowering interventions such as antihypertensive drugs or sodium restrictions. Identifying and monitoring these

DOI: 10.1201/9781003464983-4

biomarkers can lead to a better understanding of how a patient is responding to a given treatment, allowing for adjustments that optimise therapeutic efficacy and minimise adverse effects.

Predictive biomarkers are used to identify individuals more likely to respond to a certain medical product or environmental factor, while prognostic biomarkers identify the likelihood of a clinical event in patients diagnosed with a certain condition. The HER2 protein, which is overexpressed in approximately 20% of breast cancer patients, is an example of a predictive biomarker, as it is used to identify whether a patient is likely to respond to HER2-targeting therapies.[8] HER2 can also be a prognostic biomarker, as its overexpression is associated with lower survival rates. Another biomarker used for prognostics is C-reactive protein, which can help identify patients with a greater likelihood of recurrent coronary artery disease events.[9]

Finally, safety biomarkers detect or predict adverse effects due to exposure to drugs or other agents, like serum creatinine, which may be used as a safety biomarker when monitoring for nephrotoxicity in patients taking drugs that affect kidney function.[10]

Clearly, biomarkers play a central role in personalised health. A personalised approach not only improves the management of existing diseases but also contributes to the development of targeted treatments, fostering more individualised and effective therapies. As a result, patients experience better health outcomes, improved quality of life, and reduced healthcare costs due to more efficient and tailored treatment plans.

Advances in personalised medicine are also made possible thanks to new biomarkers being discovered at unprecedented speed—at the time of writing, there were over 400,000 studies listed in PubMed with the word biomarker(s) in their title or abstract. MarkerDB, a database of all known clinical and a set of pre-clinical molecular biomarkers, contains close to 28,000 entries.[11] The acceleration in the discovery of novel biomarkers has been possible thanks to innovations in technologies such as next-generation sequencing, mass spectrometry, and bioinformatics, including the use of machine learning (ML) and artificial intelligence (AI).

The elusive hormone leptin

Leptin is an adipocyte hormone, secreted by fat cells to regulate food intake and body weight.[12] Leptin is also called the "satiety hormone," as levels increase proportionally to fat mass, consequently suppressing food intake, whereas weight loss leads to a decrease in leptin levels and a concomitant increase in food intake.[13] Leptin plays a role in the pathogenesis of obesity, which is related to either leptin deficiency or resistance. In a way, leptin is to obesity what insulin is to diabetes: congenital type 1 diabetes is caused by insulin deficiency[14] in the same way as early-onset obesity is related to congenital leptin deficiency.[15] On the other hand, type 2 diabetes (T2D) is caused by hyperglycaemia and consequent insulin resistance,[16] similar to the most common forms of obesity associated with leptin resistance.[17]

Since its discovery in 1994,[18] leptin has had a wavering reputation. At first, leptin's key role in metabolism and obesity and the promising results of leptin therapy in treating severe obesity due to congenital leptin deficiency[19] gave big hopes for the therapeutic potential of leptin. However, the development of leptin treatments for obesity aggressively pursued in the late 1990s was quickly abandoned, and in 2005, leptin was defined as "a big disappointment."[20] Interest in leptin research experienced a significant decline in the early 2000s primarily due to the discovery of leptin resistance in obese individuals, which dampened the initial excitement about leptin's potential as an anti-obesity treatment. Congenital leptin deficiency is extremely rare,[21] and most cases of obesity are characterised by an overabundance of leptin and leptin resistance, rendering leptin therapies ineffective. For a while, arguably for a decade after the early 2000s, leptin's main significance was considered to be in relation to the insights it provides around the energy balance system.[22] Interest in leptin was reignited in the mid-2010s, as researchers began to uncover the role of leptin in the regulation of other physiological processes such as reproduction, immune function, and glucose metabolism, opening up the possibility of leptin as a biomarker for a variety of conditions. Given its involvement in the pathophysiology of diabetes, leptin may be a valuable prognostic marker for diabetes complications.[23] Moreover, leptin could potentially become a diagnostic and prognostic biomarker for a range of cardiovascular diseases including stroke, coronary heart disease, peripheral arterial disease, and carotid artery disease.[24] Leptin plays a central role in the reproductive system[25] opening up the possibility of using leptin as a biomarker for various applications—for example, the concentration of leptin in follicular fluid has been found to mirror ovarian reserve, making leptin a candidate biomarker to predict oocyte maturity, which is used to assess the embryo's developmental potential in in-vitro fertilisation.[26]

Leptin has recently been found to have a relationship to various mental health-related and psychiatric conditions, including stress,[27] depression,[28] suicide behaviour,[29] and Alzheimer's disease.[30] In particular, the leptin system could be a therapeutic target in Alzheimer's disease, as leptin has been shown to have neuroprotective properties.[31] Given that leptin levels are lower in people with Alzheimer's, leptin could also be a potential diagnostic and prognostic biomarker for the disease.[32,33] Finally, leptin is a promising biomarker for inflammatory bowel disease, as it may help to differentiate between Crohn's disease and ulcerative colitis.[34] Figure 4.1 summarises the various roles of leptin in human metabolism.

These recent findings suggest that future research may further prove the value of leptin as a biomarker for a variety of applications, highlighting the need to develop technologies for its quick and reliable detection. The availability of new sensing technologies can accelerate not just research efforts, but also the practical application of leptin in healthcare.

BIOMARKER DETECTION AT THE POINT OF CARE

Currently, health systems are unable to effectively care for and pay for a growing elderly population with increasingly complex and multimorbid disease,[35] highlighting the requirement for more investment in preventative strategies. While identifying and detecting biomarkers holds enormous value in the prevention and management of

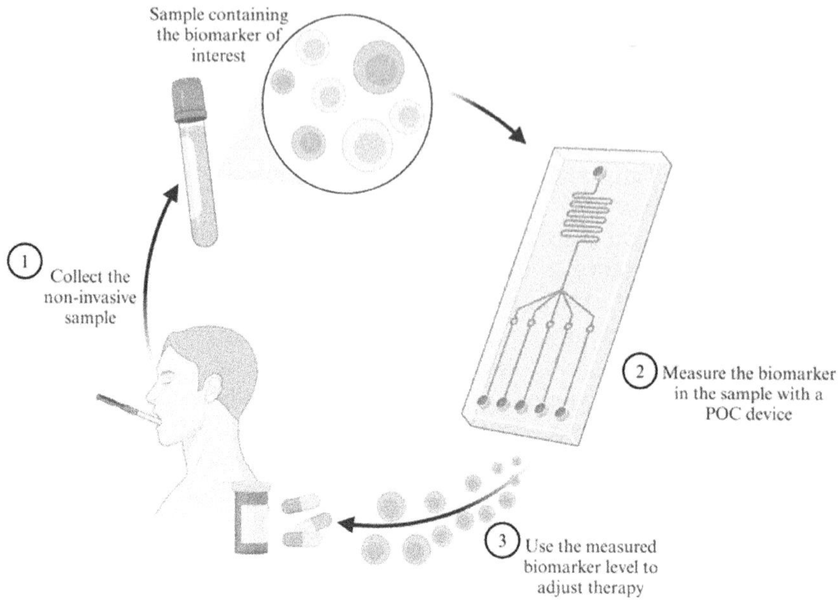

FIGURE 4.1 The hormone leptin as a biomarker of metabolic health. Created in BioRender. Cavallo, F. (2024) BioRender.com/n19g392

disease, there is a simultaneous need to ensure that biomarker detection capabilities are scalable, accessible, and equitable. In this view, point-of-care (POC) testing is a promising tool that can enable personalised healthcare by providing rapid and precise biomarker detection at the point of care, i.e., close to the patient instead of in a laboratory setting.[36] POC platforms detect and quantify biomarkers without the need for specialised laboratory facilities, hence allowing fast diagnosis and treatment,[37] and leading to better outcomes for the patient. POC devices allow portable and inexpensive tests and eliminate the need to transport samples to laboratory facilities. They can use samples like saliva or swabs that do not require trained personnel for collection. Additionally, they can detect or quantify multiple biomarkers or pathogens.[38]

POC tests are not a new concept, and tests for diabetes, anaemia, pregnancy, and HIV, among others, have become common diagnostic tools across the globe.[39] POC tests have been extensively used to detect infectious diseases, including malaria, Covid-19, Zika virus, HIV and AIDS, and Ebola virus,[40] and they are particularly cost-effective in low-resource countries. These common POC tests are what Jani et al.[41] refer to as "first-generation" POC diagnostics. First-generation tests typically use oral fluids such as saliva, urine, and blood. In some cases, like for glucose monitoring, sensing mechanisms are paired with automated readings, allowing for the quantification of the molecular target. First-generation POC tests are characterised by simple biochemical reactions often using antibodies and antigens. Lateral and vertical flow tests with visual interpretation of the result are used for target detection.

Lateral flow tests are the most used type of POC test[42]—long available for home pregnancy, drug, and alcohol testing, they have been made particularly well known during the Covid-19 pandemic. Lateral flow tests for POC have been developed for a variety of applications, including clinical (e.g., for the detection of heart attack[43] and blood clotting[44]), environmental (e.g., for pesticides[45] and bacterial monitoring[46]) and food safety applications.[47]

The great need for rapid, portable, inexpensive, and accurate POC tests drove significant progress in the field, and at the beginning of the 2010s a second generation of POC diagnostics appeared.[48] Second-generation POC tests make use of advancements in microfluidics, microelectronics, and capture and detection agent engineering to detect and quantify more complex targets with higher accuracy. The field of POC tests has seen incredible development in recent years, and significant progress has been made in regards to portable mass spectrometry,[49,50] lab-on-chip and microfluidics,[51] devices based on nanoparticles,[52] and biosensors using novel detection mechanisms. This trend was further accelerated by the Covid-19 pandemic: the urgent need for widespread testing led to the rapid development and deployment of various POC diagnostic tools. In addition to lateral flow assays, several nucleic acid-based tests for the detection of Covid-19 at the point of care have been developed,[53,54] paving the way for nucleic acid-based POC tests.

Next-generation POC tests are powered by integration with mobile and wearable technologies to provide enhanced data collection, analysis, and real-time health monitoring. Wearable sensors allow the continuous monitoring of a broad spectrum of biomarkers, and therefore, can help identify temporal patterns to better understand how biomarkers change before acute health events, such as stroke or epileptic seizures, with the aim of eventually predicting such events ahead of time. Wearable sensors measuring molecular biomarkers have been developed to measure lactate and glucose in real time,[55] to detect the melanoma biomarker tyrosinase,[56] and to monitor oestradiol.[57] Given their integration with other technologies and their real-time capabilities, next-generation POC tests could be well suited for closed-loop lifestyle interventions (Figure 4.2), in which the health status of the patient can be constantly monitored, and therapies adjusted in real-time as necessary.

The integration of sensors with AI can enhance the detection and quantification capabilities of a device.[58,59] In addition, embedding AI in POC platforms can provide predictive analytics and inform on the risk of a certain health outcome happening based on the latest biomarker reading.[60] Another characteristic of next-generation biosensors is multiplexed detection, which allows the simultaneous detection of multiple markers. The ability to detect multiple biomarkers at once is crucial when diagnosing diseases with similar symptoms and similar biomarker profiles, in addition to improving efficiency and reducing costs.

It is easy to see why accurate, inexpensive, and accessible POC tests are essential to fully realise personalised and patient-centric healthcare. Personalised medicine is inherently data hungry, as its realisation depends on the availability of patient information, and therefore, it requires the ability to measure biomarkers quickly and often. Data collection methods built on smartphones and commercially available wearable trackers are already providing a platform to collect digital biomarkers.

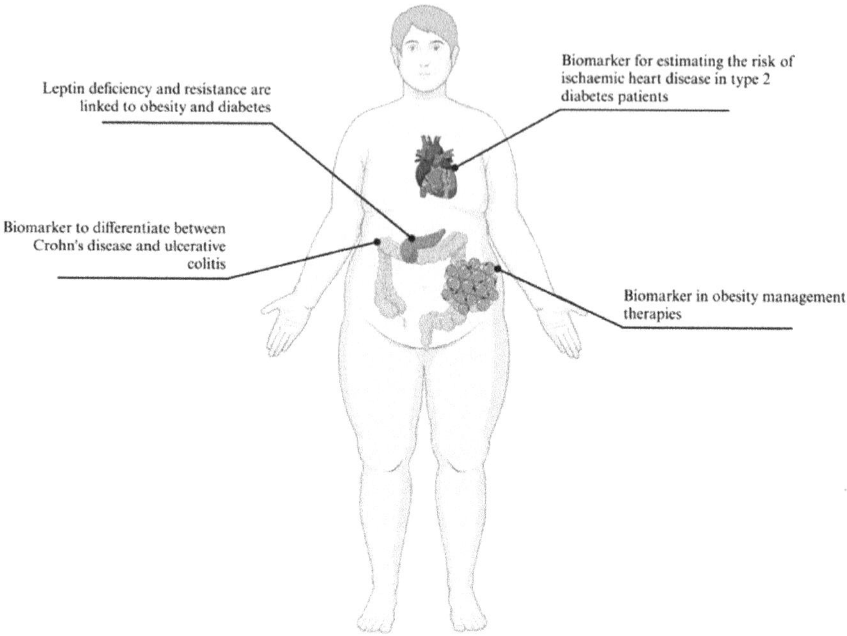

FIGURE 4.2 Example of continuous biomarker measurements at the point of care for personalised adjustment of medical therapy. Created in BioRender. Cavallo, F. (2024) BioRender .com/v15d510

POC tests are required, either at home or in community and primary care settings, to measure molecular and genetic biomarkers that can truly unlock personalised medicine. Progress is being made to implement POC tests and place them on the market, and there are active efforts to render a wider range of POC more accessible and integrated in order to enable personalised healthcare.[61]

The ability to detect and quantify biomarkers has enormous benefits not only from a personal health perspective but also for health systems. The widespread adoption of POC testing can lead to cost savings by reducing the number of hospitalisations thanks to early diagnosis and intervention.[62] In primary care, POC testing is a cost-effective alternative to laboratory tests and reduces non-attendances associated with laboratory testing.[63] Additionally, the rapid diagnosis offered by POC tests can help to reduce waiting times for results of laboratory tests, which can often take several days or even weeks.[64] As well as reducing health system inefficiencies, POC testing improves patient experience.[65] POC testing at the provider's side reduces the need for further appointments and the associated waiting times, while devices owned by the patient can help individuals feel more in control of their own health.

Challenges and outlook

Despite the promises, there are several challenges to realising the full potential of POC tests. These include technical challenges, regulatory challenges, challenges around privacy and security, as well as cost and accessibility.[66] The very nature of POC tests, which are operated outside laboratory settings by non-specialised users and often by the patients themselves, makes them prone to user error.[67] In the absence of a laboratory facility, tests are more likely to be subjected to suboptimal storage and environmental conditions, such as humidity, temperature, and light. Limitations in the accuracy and precision of POC devices are a top concern, especially for mobile and wearable tests.[68]

Although most healthcare professionals see the benefits of POC testing,[69] some scepticism remains.[70] Fighting distrust will be key in ensuring the widespread adoption of platforms needed to realise cost savings and patient benefits. Further unrealised potential lies in the integration of POC test results with other data sources, including medical history. Automatic integration with a patient's electronic health record would ensure the value of POC platforms for health providers, while also offering enhanced data for more accurate disease risk prediction models.

BIOSENSORS FOR PROTEIN BIOMARKERS

As discussed in previous sections, proteins are important biomarkers that can help diagnose and monitor various conditions, including cancer, cardiometabolic diseases, fertility issues, and infectious diseases, among many others. In the previous section, the need for POC tests for healthcare and personalised medicine was discussed, without much consideration for the technicalities of POC platforms. POC tests for biological and medical applications rely on biosensors, analytical devices that convert a biological response into an electrical signal,[71] which comprise two main components: bioreceptors to intercept the molecule being detected, and transducers to generate a measurable signal after the bioreceptor has interacted with the target.[72] This section delves deeper into biosensors for protein biomarkers, with particular emphasis on the sensing mechanism, enabled by bioreceptors, and the transducing mechanism.

The sensing mechanism can be achieved using a diverse array of bioreceptive molecules, primarily including antibodies, aptamers, enzymes, and cells or cell receptors. The detection mechanism can be implemented through various methods, with electrochemical and optical sensing being the most common techniques.[73] Biosensors can be classified based on either the sensing or the transducing technique that they employ. If classified by the bioreceptor, biosensors can be enzymatic biosensors, immunosensors, aptamer or nucleic acid-based biosensors, and whole-cell biosensors.[74] If classified based on the transduction mechanism, we have electrochemical, electronic, thermal, optical, and mass-based sensors.[75] Biosensors can also be classified based on their technology, as in the example of surface plasmon resonance (SPR) and lab-on-a-chip platforms.

Following the bioreceptor-based classification, this section discusses immunosensors and aptamer-based sensors.

IMMUNOSENSORS

Immunosensors utilise antibodies as their biorecepting molecule, which capture the target analyte thanks to their particular high affinity for it. Immunosensors are widely used for protein detection, and protein quantification is traditionally performed with the enzyme-linked immunosorbent assay (ELISA),[76] which is considered the gold standard for the determination of proteins.[77] ELISAs have been developed and are commercially available for a wide number of target molecules.[78,79,80,81] The working principle relies on antibodies that specifically bind to the target molecule. The antibodies are tagged with enzymes that develop colour upon contact with a suitable substrate, allowing the quantification of the unknown sample. Depending on the configuration, ELISAs can be direct, indirect, sandwich, or competitive.[82] Figure 4.3 shows the protocol for the commonly used sandwich ELISA assay, where each of the steps depicted is manually done for several wells and is interleaved by incubations of varying times, from 30 minutes to several hours. Despite being an established method for protein analysis in clinical practice, ELISA suffers from certain disadvantages, the major one being the laborious and time-consuming steps in high-resource settings such as specialised laboratories. Additionally, ELISA assays require several washing rounds after each step, which further increases the processing time. In summary, the nature of ELISA assays makes them challenging to adapt for POC use.

Although ELISA assays for quantification are mainly carried out in laboratories, recent attempts have been made at integrating the technique into POC platforms. For example, Liu et al.[83] developed a fully integrated chip implementing a sandwich ELISA with magnetic beads for the quantification of C-reactive protein—a marker of systemic inflammation related to CVD and metabolic syndrome[84]—in human serum. Paper-based implementations of ELISA have been, and continue to be, explored for POC applications. They have been developed for various biomarkers,

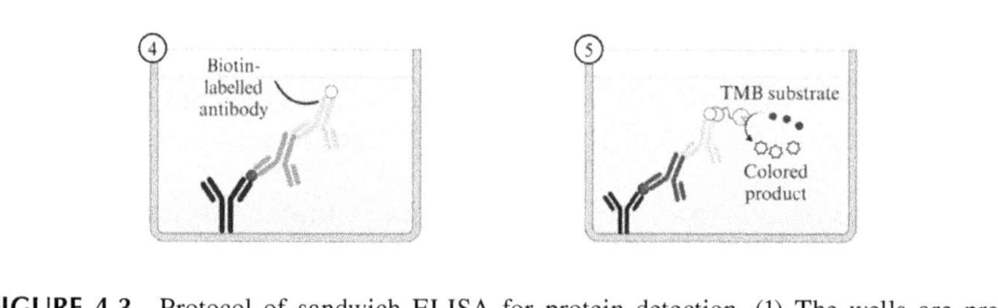

FIGURE 4.3 Protocol of sandwich ELISA for protein detection. (1) The wells are precoated with the capture antibody and the sample is added; (2) the capture antibody binds to the target molecule in the sample; (3) the primary antibody binds to the immobilised target; (4) the biotin-labelled antibody binds to the detection antibody; (5) the horseradish peroxidase (HRP)-conjugated streptavidin binds to the biotin on the detection antibody and produces a detectable colour when flushed with the TMB (3,3',5,5' tetramethylbenzidine) substrate. Created in BioRender. Cavallo, F. (2024) BioRender.com/e14z503.

such as the autoimmune disease biomarker bullous pemphigoid,[85] for the biomarker of myocardial infarction troponin,[86] and the inflammatory biomarker C-reactive protein,[87] among many others. These adaptations of ELISA for POC settings rely on the human eye for the results reading, thus significantly limiting the sensor resolution. Mobile cameras can be used to circumvent this limitation, as Murdock et al. have done for the development of a paper-based ELISA for the human performance biomarker neuropeptide Y,[88] or Pang et al. for a paper-based ELISA to detect the food-borne pathogen *E. coli* O157.[89] Additionally, ELISAs have been automated through integration with microfluidic devices[90,91] and lab-on-chip devices[92,93,94] making the assays more suitable for POC applications.

As an alternative to optical immunosensors, assays and sensors that detect electrochemical changes in samples, also known as electrochemical biosensors, are a promising option for POC protein quantification, since they more easily allow for miniaturisation, portability, rapid detection speed, and on-site monitoring capabilities.[95] Additionally, electrochemical biosensors allow the development of multiplexed devices able to combine assays for different molecular targets. These can be realised in various formats, for example, wearable and implantable devices, tattoos, and contact lenses.[96]

In terms of transducing mechanisms, electrochemical immunosensors have been paired with chronoamperometry, differential pulse voltammetry, square wave voltammetry, and electrochemical impedance spectroscopy.[97,98] Recent research in electrochemical biosensors has worked on improving limits of detection in order to achieve high sensitivity,[99] research that benefited greatly from advancements in nanotechnologies and nanomaterials. Sensor sensitivity, as well as selectivity, can be enhanced through modification with nanomaterials,[100] such as metallic nanoparticles which improve transfer efficiency and increase the electrode's surface area, thus leading to signal amplification.[101,102] Other nanomaterials commonly used to enhance electrochemical biosensors include carbon nanotubes, graphene, and nanowires.[103] Electrochemical biosensors have been developed for a wide range of protein biomarkers, including biomarkers of cancer, viral infectious diseases, and markers of inflammation.[104]

Other immunoassays employing different kinds of transducers exist, such as piezoelectric and magnetic immunoassays. Piezoelectric sensors, which generate an electrical current in response to mechanical stress, allow for label-free detection and are strong candidates for a wide variety of applications due to their cost-effectiveness, experimental simplicity, and real-time output.[105] Magnetic immunoassays are a promising alternative for protein detection applications and rely on magnetic nanoparticles that can easily be combined with existing sandwich immunoassay techniques to create highly specific and sensitive sensors.[106]

Aptamer-based biosensors

Antibodies are a very established choice of bioreceptor and have been employed in countless biosensors, including for POC applications. Even though existing immunosensors can achieve low limits of detection with low sample volumes, and in many cases with low times to result, antibodies present limitations related to their stability, shelf life, and costs.[107,108] In the last decade, aptamers have shown promise as an alternative to antibodies for biosensing applications. Aptamers are short single-strand sequences of DNA/RNA that are selected in vitro through systematic evolution of ligands by exponential enrichment (SELEX).[109,110] During SELEX, the target-specific aptamer is selected from a large sequence library based on affinity to the target molecule, allowing to reach very high affinity levels. The high affinity can be attributed to the secondary and tertiary structures that nucleotide sequences can form, as aptamers function through shape-specific recognition of binding partners.

Aptamers are a popular alternative to antibodies due to several desirable characteristics, such as being more chemically and thermally stable, reproducible, and cost-effective.[111]

The shelf life of aptamers is longer compared to the shelf life of antibodies, even in unfavourable environmental conditions,[112] making them particularly suited for POC applications and use in low-resource settings. Additionally, aptamers can be modified post-SELEX through various techniques that improve the sequences' properties,[113] and they can be regenerated to allow reusable biosensors.[114]

As an alternative bioreceptor to antibodies, aptamers can be used in the same configurations as antibodies and paired with optical or electrochemical transducers. Aptamers can be used for in-vitro assays and can be integrated into POC sensing platforms such as lateral flow assays, paper-based assays, lab-on-a-chip, and wearable devices. Aptamers have been extensively used in lateral flow assays,[115] and have been developed for use in low-resource settings for a variety of applications: food safety,[116] drug abuse testing,[117] pregnancy testing,[118] detection of cancer biomarkers,[119] and infectious disease diagnosis.[120] Aptamers have also been integrated into paper-based aptamer sensors, for example, for multiplexed detection of cancer biomarkers,[121] and in lab-on-a-chip devices to monitor of HbA1c,[122] to detect cancer biomarkers,[123] and for food safety applications.[124] More recently, aptamers have been used in wearable devices, and aptamer-based wearable sensors have been developed to continuously monitor oestradiol[125] and cortisol.[126]

A unique feature of DNA/RNA-based aptamers is the possibility of quantification through real-time, or quantitative, polymerase chain reaction (qPCR), a robust and sensitive DNA quantification method with limits of detection of a few sequences that significantly improve sensitivity and resolution.[127] Aptamers combined with qPCR amplification have been previously used for various detection purposes. For example, Guo et al.[128] developed an ultrasensitive assay based on aptamers and qPCR amplification for the food toxin aflatoxin B1 that achieved extremely high sensitivity (25 fg/mL). Aptasensors with immobilised aptamers and qPCR have also been developed for ochratoxin A, another food contaminant.[129] Other DNA amplification methods, such as rolling cycle amplification, have also been implemented for the detection of the food contaminant Vibrio parahaemolyticus[130] and for quantifying adenosine triphosphate.[131] qPCR-enhanced aptasensors also have medical applications, and have

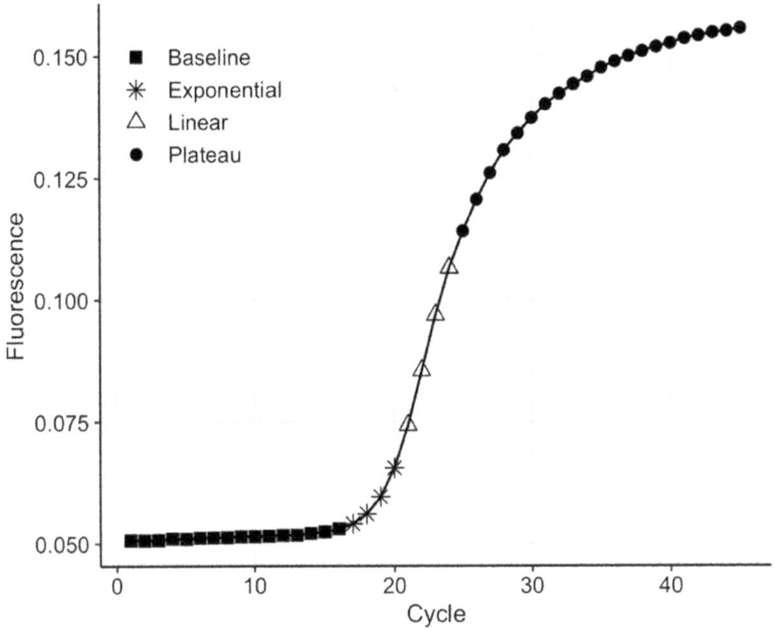

FIGURE 4.4 The four phases of the PCR amplification curve. Reprinted with permission from Cavallo F. R., Mirza K. B., de Mateo S., Miglietta L., Rodriguez-Manzano J., Nikolic K., Toumazou C. "A Point-of-Care Device for Fully Automated, Fast and Sensitive Protein Quantification via qPCR." *Biosensors*. 2022; 12(7):537. https://doi.org/10.3390/bios12070537

been shown to work well for the quantification of proteins such as platelet-derived growth factor and thrombin in serum and plasma samples.[132]

ALGORITHMS FOR DNA QUANTIFICATION THROUGH QPCR

The performance of biosensors employing nucleic acid aptamers can be significantly enhanced through the use of qPCR. qPCR is based on PCR, a technique to amplify DNA sequences. One PCR cycle is composed of three steps, carried out at different temperature ranges: denaturation, during which the double-stranded DNA is separated into single strands; primer binding, during which the primers, small DNA sequences designed to bind to the target DNA, anneal to the single-strand DNA; and DNA synthesis, during which the primers extend the single-strand DNA. PCR cycles can be repeated many times, resulting in an exponential accumulation of the target DNA.

qPCR is an extension of PCR in which the process of amplification is measured in real time using fluorescent dyes and optical detection.[133] During qPCR, an amplification curve is built, which is characterised by four distinct phases (Figure 4.4). During the background phase, the target DNA is too low in concentration, and the background noise covers the signal. Once the target DNA becomes detectable, the output signal enters the exponential phase, in which the DNA is doubled at every cycle. In the exponential region, the curve is described by the equation $y_n = y_0(1+E)^n$, where n is the cycle number, y_n is the number of DNA copies at cycle n, y_0 is the initial amount of DNA, and E is the reaction efficiency.[134] The linear phase appears due to inefficiencies in the amplification process, such as suboptimal melting temperatures for the primers and non-specific amplification. Finally, the process enters a plateau due to the depletion of qPCR resources, such as polymerase and primers.

After completion of the qPCR cycles, several algorithms are employed to infer the original quantity of target DNA from the full amplification curve. Before delving into the different algorithms, a few concepts need to be introduced. Firstly, the threshold cycle (C_t) value is the cycle number at which the fluorescence generated within a reaction crosses the fluorescence threshold, a fluorescent signal significantly above the background fluorescence. This value is inversely proportional to the DNA concentration. In the most common way to quantify DNA, a standard curve is built by including wells with known DNA concentrations, called templates, in the qPCR plate. The resulting C_t values of the templates generate a standard curve, which is used to quantify the concentration of the targeted substance. In order to obtain a reliable standard curve, it is crucial to obtain precise C_t values, precision that is hindered by poor quality templates and the presence of inhibitors found in biological samples.[135]

To obtain the C_t value, several different methods can be employed. The simplest one, named the “fit point” or “threshold cycle” method, involves setting the C_t as the cycle at which the fluorescence signal intersects a pre-set threshold.[136] This approach is sensitive to scale and, therefore, requires scaling of the data, which is usually implemented by fitting a logistic function.[137] Other methods look at the first or second derivatives of the fluorescent curve, and the cycle at which they reach the maximum is set to be the threshold C_t. Other algorithms also exist, and they can achieve more accurate C_t estimates. For example, Zhao et al. use a combination of functions to model the various phases of the amplification curve, which are then used to determine the C_t value and the reaction efficiency required to estimate the target concentration.[138] Guescini et al. firstly fit a Richard’s function, an extension of the logistic growth curve, to the amplification curve, and use the estimated parameters to calculate a C_t value that accounts for the reaction kinetics.[139] The C_t values can also be derived in real-time during the qPCR amplification by estimating a model at each cycle until a threshold is reached, after which the qPCR is interrupted. Han et al.[140] have demonstrated that this method can achieve high accuracy and significantly reduce the number of cycles needed for quantification.

More recently, ML methods have been employed to improve the quantification process. Moniri et al.[141] combined the aforementioned methods to generate a multidimensional standard curve that optimises qPCR performance by employing multiple features and automatically detecting outliers. Alouani et al.[142] used Convolutional Neural Networks to analyse qPCR fluorescence curves as a whole image to improve the diagnosis of SARS-CoV-2.

CASE STUDY: APTASENSOR FOR THE QUANTIFICATION OF LEPTIN THROUGH PCR

This case study discusses aptasensors for the quantification of leptin using qPCR, which allows for achieving very low limits of detection. To detect leptin in serum and plasma, ELISA is the most widely used method.[143] Immunosensors for leptin have been developed using optical[144,145] and electrochemical transducing techniq ues.[146,147,148,149,150,151,152,153] For example, He et al.[154] detected leptin with a chemiluminescent immunoassay fabricated with nanoparticles and achieved very low limits of detection (0.3 pg/mL) with a low sample volume (100 μL); however, the fabrication process takes more than 40 hours, and the incubation time with leptin takes 2 hours. The electrochemical immunosensors developed by Dong et al.,[155] Cai et al.,[156] and Chen et al.[157] achieved limits of detection in the pg/mL, but with long times to result. Ojeda et al.[158] developed an immunosensor for leptin with a 95-minute time to result.

As discussed in previous sections, aptamers provide a very promising alternative to antibodies. Despite a leptin aptamer being synthesised in 2013,[159] no aptasensor was developed for leptin until 2021.[160] Since then, other aptasensors for the determination of leptin have been developed, such as a label-free impedimetric aptasensor based on nanoparticles,[161] and a sandwich-type voltametric aptasensor.[162]

This case study discusses the development of the first aptasensor employing PCR amplification to detect leptin. Despite being validated on leptin, this aptasensor technique can be employed to detect any protein target, provided a suitable aptamer exists.

Methods: Reagents and sequences

Leptin was purchased from Thermo-Fisher as part of the Invitrogen Leptin Human ELISA Kit (Catalog no. KAC2281); lysozyme was also purchased from Thermo-Fisher. A leptin binding buffer (50 mM Tris pH 8.0, 140 mM NaCl, 1 mM MgC12 (pH 7.5)) was used in the sample incubation step. The chemicals used to prepare the buffers were purchased from Sigma-Aldrich.

The oligonucleotide sequences were purchased from IDT DNA Technologies. Primers were synthesised with standard desalting, while the leptin adaptor with high-performance liquid chromatography purification. The leptin aptamer was taken from Ashley and Li.[163] The adaptor was designed to have 100bp from the Green Fluorescent Protein (GFP) gene with an extra 26bp for aptamer binding, while the complex sequence was designed to contain the forward adaptor sequence and the reverse aptamer sequence. The adaptor forward and reverse primers were taken from Neuzil et al.[164] All the oligonucleotide sequences are shown in Table 4.1.

For qPCR, the SYBR Green FastStart® Essential DNA Green Master from Roche LifeScience™ was used.

Methods: Protocol with beads

Initially, the aptasensor was developed with the use of magnetic beads (Figure 4.5). Firstly, the beads were washed with 500 μL of Buffer 1 (20 mM Tris pH 8,

TABLE 4.1
Oligonucleotide sequences used in the assay. Primers 1 and 2 are used for amplification of the adaptors, Primers 3 and 4 for the amplification of the complexes

Name	Sequence
Aptamer	GTTAATGGGGATCTCGCGGCCGTTCTTGTTGCTTATACA
Adaptor	GCTACCCCGACCACATGAAGCAGCACGACTTCTTCAAGTCCGCCATGCCC GAAGGCTACGTCCAGGAGCGCACCATCTTCTTCAAGGACGACGGCAACTA AAAAATGTATAAGCAACAAGAACGGC
Complex	GCTACCCCGACCACATGAAGCAGCACGACTTCTTCAAGTCCGCCATGCCC GAAGGCTACGTCCAGGAGCGCACCATCTTCTTCAAGGACGACGGCAACTA AAAAATGTATAAGCAACAAGAACGGCCGCGAGATCCCCCATTAAC
Primer 1	CACATGAAGCAGCACGACTT
Primer 2	CGTCGTCCTTGAAGAAGATGGT
Primer 3	CACATGAAGCAGCACGACTT
Primer 4	TGGGGGATCTCGTGGC

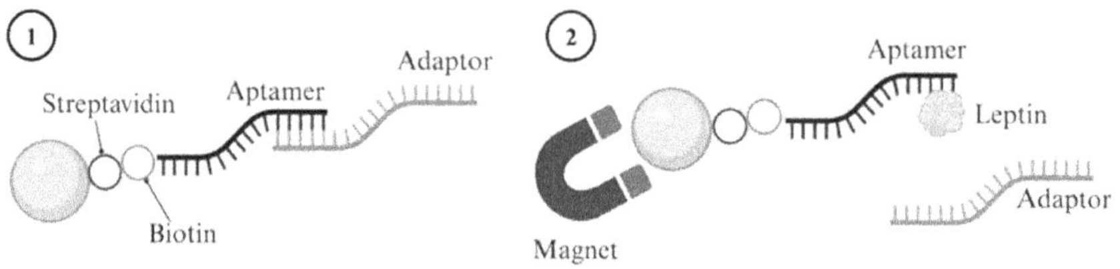

FIGURE 4.5 Proposed method using magnetic beads. (1) The aptamer-adaptor complex is formed on the magnetic bead through binding of biotin and streptavidin. (2) Given the high affinity of the aptamer for leptin, the adaptor is displaced from the complex as leptin binds to the aptamer. Adaptors—proportional to leptin concentration—are then amplified by PCR. Created in BioRender. Cavallo, F. (2024) BioRender.com/s85q706. Reprinted with permission from F. R. Cavallo, K. B. Mirza, S. de Mateo, K. Nikolic, J. Rodriguez-Manzano, and C. Toumazou, "Aptasensor for Quantification of Leptin Through PCR Amplification of Short DNA-Aptamers." *ACS Sensors* 2021 6 (3), 709-715. DOI: 10.1021/acssensors.0c02605.

150 mM NaCl, 5 mM MgCl2) three times. In the last wash, the beads were resuspended with 1 L of Buffer 1. Then, 5 μL of 100 μL aptamers were added to the vial containing the beads, gently mixed, and incubated at 37°C for 15 minutes. With the aid of a magnet, the beads were washed with 500 μL of leptin binding buffer (three times) to remove the unbound aptamers. The beads were resuspended in 1,000 μL of binding buffer.

Secondly, the adaptors were placed at 95°C for 5 minutes to disrupt any secondary structure. Subsequently, 5 μL from the 0.1 nM stock was added in the vial containing the aptamer-coated beads and incubated at 37°C for 15 minutes. Using a magnet to capture the beads, the vial was washed three times with the buffer (500 μL per wash) to remove the unbound sequences. The supernatant from the washes was kept to later quantify the amount of aptamers released during this step. Aptamer quantification is possible by amplifying the complexes formed between free adaptors and aptamers; the process is explained in detail in the next section.

Finally, in a new tube, 10 μL of leptin and 80 μL of buffer were added to 10 μL of prepared beads and incubated at 37°C for 15 minutes. The vial was centrifuged at 10,000 × g for 1 minute and the supernatant was collected for qPCR analysis.

qPCR was done using the SsoFast EvaGreen® supermix from Bio-Rad™ in a 20 μL reaction volume consisting of 10 μL of supermix, 2 μL of forward and reverse primers (5 μM each), 5 μL of DNA/RNA-free water, and 3 μL of DNA template from the supernatant collected after centrifugation.

The qPCR cycling conditions were as follows: enzyme activation at 98°C for 2 minutes, denaturation at 98°C for 5 seconds, and annealing at 60°C for 20 seconds (for 40 cycles). Fluorescence measurements were taken after each annealing step.

Methods: Protocol without beads

During the optimisation of the assay discussed in the previous section, the assay was further simplified, and the final method is illustrated in Figure 4.6. Adaptors were incubated at 95°C for 5 minutes to disrupt the secondary structures and then mixed with aptamers in a 1:1 (v/v) ratio. After mixing gently, they were incubated at 37°C for 15 minutes to generate the aptamer-adaptor complex by hybridisation. The adaptor-aptamer complexes were mixed with leptin and leptin binding buffer in a 1:1:8 (v/v) ratio and incubated at 37°C for 30 minutes (Figure 4.6(1)). This protocol relies on the

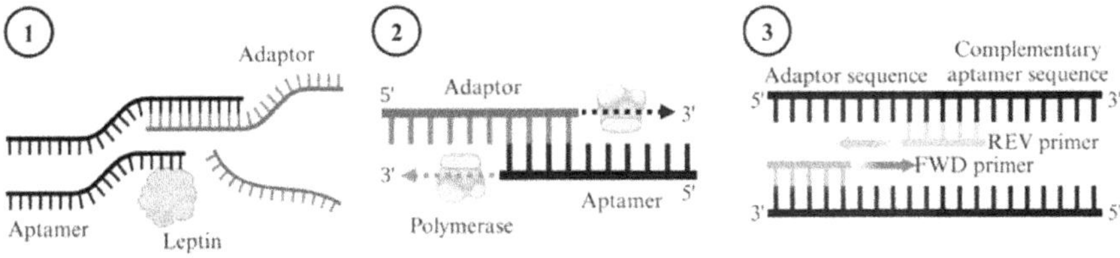

FIGURE 4.6 Proposed method for aptamer-based protein quantification. (1) As the affinity of the aptamer is higher for leptin than for the adaptor, the adaptor is displaced from the complex as leptin binds to the aptamer. (2) The complexes not displaced by leptin are incubated with polymerase enzyme to elongate the complexes and create the target DNA for PCR amplification. (3) The elongated complexes—inversely proportional to leptin concentration—are amplified with a forward primer binding to the adaptor part of the elongated complex, and a reverse primer binding to the aptamer part of the complex. Created in BioRender. Cavallo, F. (2024) BioRender.com/i57a507. Reprinted with permission from F. R. Cavallo, K. B. Mirza, S. de Mateo, K. Nikolic, J. Rodriguez-Manzano, and C. Toumazou, "Aptasensor for Quantification of Leptin Through PCR Amplification of Short DNA-Aptamers." *ACS Sensors* 2021 6 (3), 709-715. DOI: 10.1021/acssensors.0c02605.

amplification of the aptamer-adaptor complexes elongated by polymerase (Figure 4.6 b), which is inversely proportional to the leptin concentration. After incubation with leptin, 5 μL of DNA template were mixed with 10 μL of FastStart Essential DNA Green Master and 3 μL DNA/RNA-free water and incubated at 60°C for 10 minutes to allow the elongation of the sequences forming the complexes (Figure 4.6(2)b). Finally, 2 μL of forward and reverse primer mix at 5 μM (10×) were added to obtain a 20 μL reaction. The forward primer anneals to the adaptor part of the complex, while the reverse primer anneals to the aptamer part, which allows for complex amplification without amplifying free adaptors (Figure 4.6(3)c). The PCR cycling conditions were pre-incubation at 95°C for 600 s and 45 cycles of three-step amplification at 95°C for 10 s, at 60°C for 10 s, and 72°C for 10 s. Fluorescence measurements were taken after each annealing step.

Methods: Aptasensor characterisation and optimisation

The optimal primer amount for the PCR was found by titrating primer concentrations between 0.3 μM and 0.7 μM. The concentrations of aptamers and adaptors were optimised by varying the concentrations from 0.33 pM to 330 nM for both reactants. The optimal concentrations were chosen to maximise the complex formation while minimising the number of free aptamers.

The impact of incubation conditions was assessed by varying the incubation time (from 10 minutes to 60 minutes) and performing multiple rounds of incubation.

To assess the impact of buffers on the reactions, three buffers were tested: the leptin binding buffer, the TGK buffer (25 mM Tris pH 8, 192 mM glycine, 5 mM K_2HPO_4 from Ashley and Li[165]) and Guo buffer (10 mM Tris, 120 mM NaCl, 5 mM KCl, 20 mM $CaCl_2$ pH 7.0 from Guo et al.[166]). For the leptin binding buffer, several values of pH were tested, from 2.5 to 8.3. The pH was adjusted by varying the concentration of HCl.

Methods: Dissociation constant calculation

For a general reaction in which $xA + yB \rightleftharpoons A_xB_y$, where x, y are the number of molecules for reactant A and B respectively, the dissociation constant K_d is defined as:

$$K_d = \frac{[A]^x [B]^y}{[A_xB_y]} \qquad (4.1)$$

To determine K_d, the binding curve was used in which the fraction of $[B]$ bound at equilibrium is plotted against the concentration of free $[A]$. To plot the binding curve, the concentration of $[A]$ is varied while holding $[B]$ fixed. The fraction of $[B]$ bound can be written as:

$$\left[B_{bound}\right]=\frac{[AB]}{[B]+[AB]}=\frac{[A]}{K_d+[A]} \tag{4.2}$$

Meaning that when $[A]=K_d$, then $\left[B_{bound}\right]=0.5$. As such, it is possible to estimate K_d as the value of $[A]$ on the binding curve at which $\left[B_{bound}\right]=0.5$.

To calculate the K_d between aptamers and adaptor, a fixed concentration of adaptors (3.3 nM) was mixed with aptamers of concentrations varying from 0.05 nM to 100 nM. To ensure the reaction reached equilibrium, the sequences were incubated for 6 hours.

Methods: Specificity and repeatability

As specificity is an important issue in aptasensors,[167] the specificity of the assay was evaluated by comparing the response to leptin and to lysozyme. To compare the aptasensor's specificity to leptin, leptin was added in varying concentrations (6.25, 62.5, 625 pM) with a fixed concentration of lysozyme (62.5 pM), and in a second experiment varying concentrations of lysozyme (6.25, 62.5, 625 pM) were added with fixed leptin (62.5 pM).

Repeatability was evaluated by pooling the C_t [168] values of the leptin standards across experiments and calculating the mean and standard error.

Methods: Assay validation with human saliva

The assay was validated on human saliva. Samples were centrifuged at 15,000 × g for 10 minutes to remove debris. To remove nucleic acids which could interfere with the assay, the sample was prepared following Xuan et al.[169]: 5% polyethyleneimine was added to the supernatant to a concentration of 0.5% (w/v), followed by incubation for 10 minutes and centrifugation at 15,000 × g for 10 minutes, 474 g/L of ammonium sulphate was added to the supernatant with stirring for 20 minutes. After centrifugation at 17,000 × g for 20 minutes, the pellet was washed three times with ammonium sulphate at 70% to remove excess polyethyleneimine. The pellet was resuspended in DNA/RNA-free water. Assay specificity was assessed by comparing the complex displacement between leptin and lysozyme, a protein present in high concentrations (μg/mL) in saliva.[170] To measure recovery, saliva samples were spiked with different concentrations of synthetic leptin (0.1 ng/mL, 1 ng/mL, 10 ng/mL, and 100 ng/mL). Recovery was calculated as:

$$\frac{C_{observed}-C_{endogenous}}{C_{spiked}} \tag{4.3}$$

where C stands for concentration.

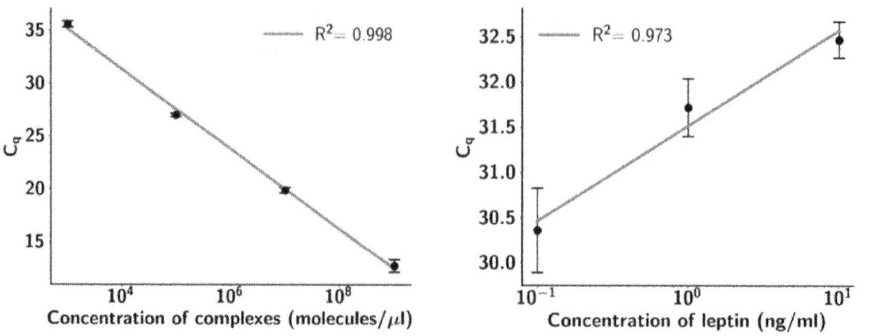

FIGURE 4.7 Standard curves for the quantification of aptamer-adaptor complexes (left) and leptin (right). Reprinted with permission from F. R. Cavallo, K. B. Mirza, S. de Mateo, K. Nikolic, J. Rodriguez-Manzano, and C. Toumazou, "Aptasensor for Quantification of Leptin Through PCR Amplification of Short DNA-Aptamers." *ACS Sensors* 2021 6 (3), 709-715. DOI: 10.1021/acssensors.0c02605.

Results

In this section, only the detailed characterisation of the assay without beads is presented, as it is the final choice of assay. Firstly, standard curves were built for the adaptor-aptamer complexes (Figure 4.7), which are also used to calculate the leptin concentration from standards during assay optimisation. Aptamers and adaptors were incubated at 37°C for 15 minutes and the formed complexes were incubated with leptin in concentrations of 10 ng/mL, 1 ng/mL, and 0.1 ng/mL. Amplification curves for leptin concentrations are shown in Figure 4.8.

Results: Assay optimisation

The optimal primer amount of 0.5 μM was found by titrating primer concentrations between 0.3 and 0.7 μM.

Aptamer-adaptor concentrations were optimised as the complex formation efficiency greatly affects the performance of the assay, as it determines the amount of complexes available for leptin displacement. The concentration of aptamer and adaptor was optimised by testing the 16 combinations of 330 nM, 3.3 nM, 330 pM, and 3.3 pM. When a ratio greater than 2:1 (adaptors:aptamers) is maintained, all aptamers form complexes; since the amount of complexes is proportional to the amount of aptamer and adaptor, the optimal concentration was chosen as 330 nM and 3.3 nM for adaptors and aptamers, respectively. Detailed results are presented in Table 4.2.

The choice of buffer influences how well complexes are formed and retained. Water, TGK, and Guo buffer were tested. The TGK buffer and distilled water produced the best results in terms of complex retention, but TGK achieved higher fluorescence (see Figure 4.9). The choice of buffer also influences the affinity between leptin and the aptamer. Figure 4.10 shows the effect of the leptin binding buffer, TGK, and the buffer in a commercially available ELISA kit on the leptin standards. The leptin binding buffer was chosen both for the aptamer-adaptor binding step and

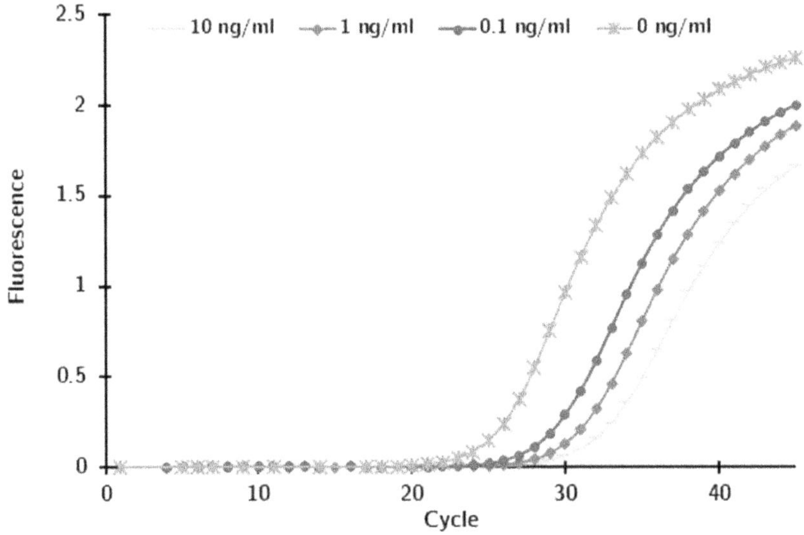

FIGURE 4.8 Amplification of leptin standards in concentrations of 10, 1, and 0.1 ng/mL. Reprinted with permission from F. R. Cavallo, K. B. Mirza, S. de Mateo, K. Nikolic, J. Rodriguez-Manzano, and C. Toumazou, "Aptasensor for Quantification of Leptin Through PCR Amplification of Short DNA-Aptamers." *ACS Sensors* 2021 6 (3), 709-715. DOI: 10.1021/acssensors.0c02605.

TABLE 4.2
Complex formation for different adaptor and aptamer concentrations

Adaptor concentration (molec/5 μL)	Aptamer concentration (molec/5 μL)	C_t	Complexes/μL	Ideal nr of complexes/μL	Free adaptors/μL	Free aptamers/μL
10^{10}	10^{12}	12	$1.18 \cdot 10^{9}$	$2 \cdot 10^{9}$	$8.24 \cdot 10^{8}$	$1.98 \cdot 10^{11}$
10^{8}	10^{12}	19.7	$9.38 \cdot 10^{6}$	$2 \cdot 10^{7}$	$1.06 \cdot 10^{7}$	$2 \cdot 10^{11}$
10^{6}	10^{12}	22.6	$1.50 \cdot 10^{6}$	$2 \cdot 10^{5}$	$-1.30 \cdot 10^{6}$	$2 \cdot 10^{11}$
10^{12}	10^{10}	11	$2.22 \cdot 10^{9}$	$2 \cdot 10^{9}$	$1.98 \cdot 10^{11}$	0
10^{10}	10^{10}	18	$2.62 \cdot 10^{7}$	$2 \cdot 10^{9}$	$1.97 \cdot 10^{9}$	0
10^{8}	10^{10}	24.8	$3.64 \cdot 10^{5}$	$2 \cdot 10^{7}$	$1.96 \cdot 10^{7}$	$1.98 \cdot 10^{9}$
10^{6}	10^{10}	31.1	$6.52 \cdot 10^{3}$	$2 \cdot 10^{5}$	$1.93 \cdot 10^{5}$	$2 \cdot 10^{9}$
10^{12}	10^{8}	19.1	$1.30 \cdot 10^{7}$	$2 \cdot 10^{7}$	$2 \cdot 10^{11}$	0
10^{10}	10^{8}	26.5	$1.25 \cdot 10^{5}$	$2 \cdot 10^{7}$	$2 \cdot 10^{9}$	0
10^{8}	10^{8}	33.4	$1.52 \cdot 10^{3}$	$2 \cdot 10^{7}$	$2 \cdot 10^{7}$	0
10^{6}	10^{8}	37.3	$1.29 \cdot 10^{2}$	$2 \cdot 10^{5}$	$2 \cdot 10^{5}$	$1.98 \cdot 10^{7}$
10^{12}	10^{6}	26.6	$1.18 \cdot 10^{5}$	$2 \cdot 10^{5}$	$2 \cdot 10^{11}$	0
10^{10}	10^{6}	32.6	$2.67 \cdot 10^{3}$	$2 \cdot 10^{5}$	$2 \cdot 10^{9}$	0
10^{8}	10^{6}	40	23.3	$2 \cdot 10^{5}$	$2 \cdot 10^{7}$	0
10^{6}	10^{6}	46.4	0.412	$2 \cdot 10^{5}$	$2 \cdot 10^{5}$	0

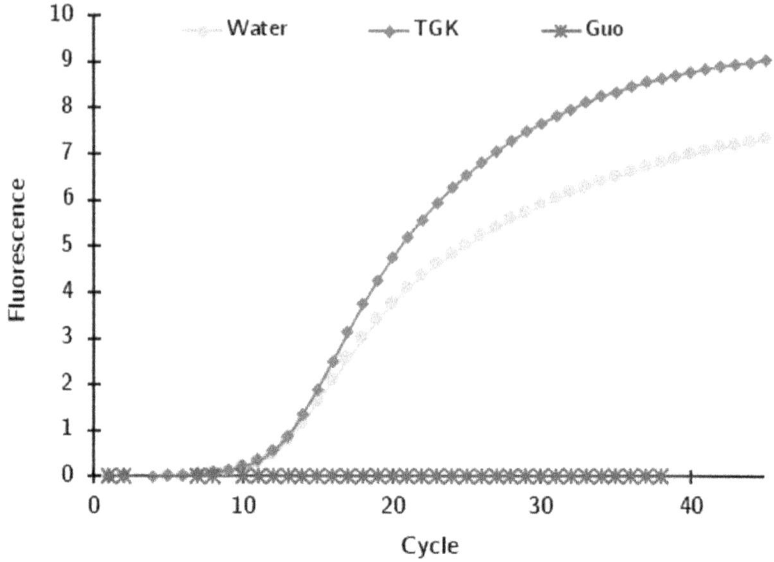

FIGURE 4.9 Effect of different buffers on complex formation. Reprinted with permission from F. R. Cavallo, K. B. Mirza, S. de Mateo, K. Nikolic, J. Rodriguez-Manzano, and C. Toumazou, "Aptasensor for Quantification of Leptin through PCR Amplification of Short DNA-Aptamers." *ACS Sensors* 2021 6 (3), 709-715. DOI: 10.1021/acssensors.0c02605.

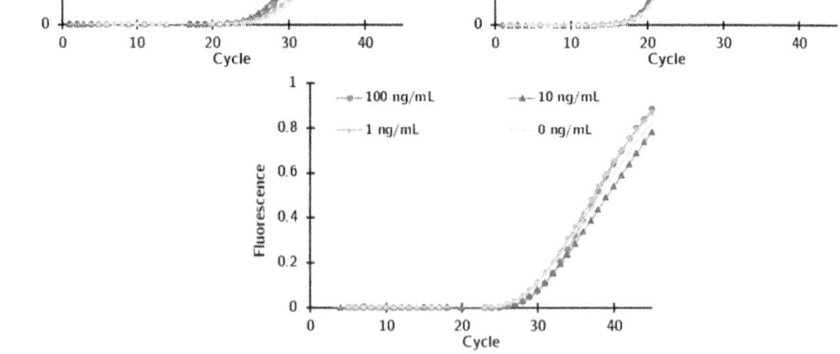

FIGURE 4.10 Effect of different buffers on leptin standards. From the top left corner, anti-clockwise: leptin binding buffer, TGK buffer, buffer 1 from the ELISA assay. Reprinted with permission from F. R. Cavallo, K. B. Mirza, S. de Mateo, K. Nikolic, J. Rodriguez-Manzano, and C. Toumazou, "Aptasensor for Quantification of Leptin Through PCR Amplification of Short DNA-Aptamers." *ACS Sensors* 2021 6 (3), 709-715. DOI: 10.1021/acssensors.0c02605.

TABLE 4.3
Value of C_t for leptin standards depending on pH of the leptin binding buffer

Leptin concentration	pH 2.5	pH 3	pH 4	pH 5	pH 5.5	pH 7.2	pH 7.7	pH 8	pH 8.3
10 ng/ml	32.86	32.47	28.51	29.67	27.68	30	27.66	28.6	32.92
1 ng/ml	31.05	31.72	28.34	31.91	26.89	29.49	28.88	29.02	32.05
0.1 ng/ml	29.35	30.36	31.33	31.23	28.23	26.91	28.2	29.43	29.43

for the leptin binding step, as it improves the displacement of complexes by leptin and exhibited higher repeatability across experiments. The pH of the leptin binding buffer influenced the ability of leptin to displace the complexes. Nine pH values from 2.5 to 8.3 were tested, with 2.5 and 3.0 producing better linearity and separation between the standard concentrations. *C*t values obtained with buffers at different pH for different standard concentrations are shown in Table 4.3.

To increase complex formation, the effect of incubation rounds and time was explored. Two rounds of incubation were tested to verify if there were any improvements in the hybridisation between adaptors and aptamers. The first round of incubation was conducted between aptamers and adaptors at 37°C for 15 minutes. For the second round, an aliquot from Round 1 and an aliquot of adaptors were mixed in equal volumes and incubated again at 37°C for 15 minutes. The second round only slightly increased the number of complexes formed and thus, was not deemed necessary. Incubation times did not affect the performance of the assay: aptamer, adaptor, and leptin were incubated with TGK for 10, 30, 45, and 60 minutes at 37°C, and the number of complexes was constant regardless of the incubation time.

Results: Assay characterisation

The inter-assay variability is shown in Figure 4.11. The specificity of the assay was assessed through interference with lysozyme, which was chosen as an interfering protein given its high concentrations in human saliva. As shown in Figure 4.12, for fixed leptin concentration (62.5 pM), the assay did not respond to varying concentrations of lysozyme, as the variation in fluorescence from the reference point is significantly lower than for varying concentrations of leptin with a fixed lysozyme concentration (62.5 pM).

The dissociation constant K_d = 14.3 nM was found using the binding curve shown in Figure 4.13. The varying aptamer concentrations were log-transformed and plotted on the x-axis, while the normalised fraction of bound aptamers was plotted on the y-axis. The amount of bound aptamers was calculated as the number of complexes formed (calculated using the standard curve for complexes) over the initial number of adaptors.

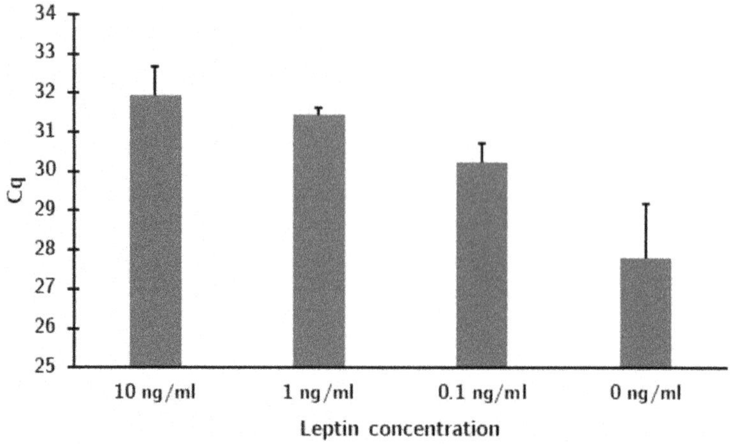

FIGURE 4.11 Repeatability of assay for leptin standards. The error bars represent the standard error. Reprinted with permission from F. R. Cavallo, K. B. Mirza, S. de Mateo, K. Nikolic, J. Rodriguez-Manzano, and C. Toumazou, "Aptasensor for Quantification of Leptin Through PCR Amplification of Short DNA-Aptamers." *ACS Sensors* 2021 6 (3), 709-715. DOI: 10.1021/acssensors.0c02605.

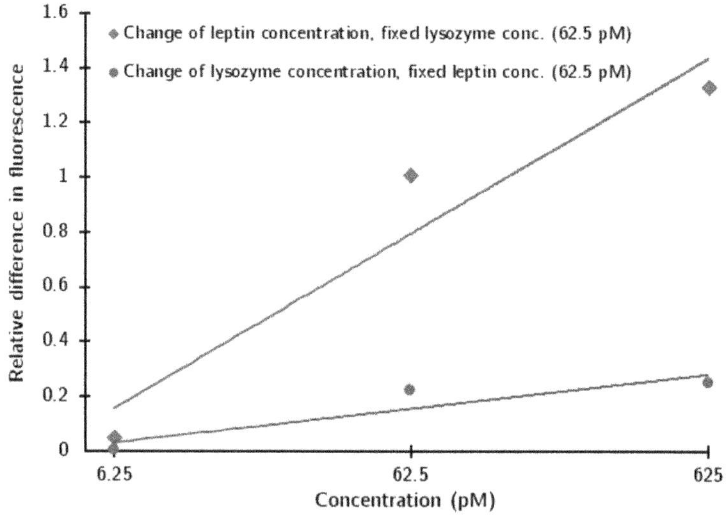

FIGURE 4.12 Comparison of the response of the assay to varying concentrations of leptin and lysozyme. Leptin was added in varying concentrations (6.25, 62.5, 625 pM) with a fixed concentration of lysozyme (62.5 pM), and in a second experiment varying concentrations of lysozyme (6.25, 62.5, 625 pM) were added with fixed leptin (62.5 pM). The reference points at 6.25 pM have been shifted slightly apart along the y-axis for better visibility. Reprinted with permission from F. R. Cavallo, K. B. Mirza, S. de Mateo, K. Nikolic, J. Rodriguez-Manzano, and C. Toumazou, "Aptasensor for Quantification of Leptin Through PCR Amplification of Short DNA-Aptamers." *ACS Sensors* 2021 6 (3), 709-715. DOI: 10.1021/acssensors.0c02605.

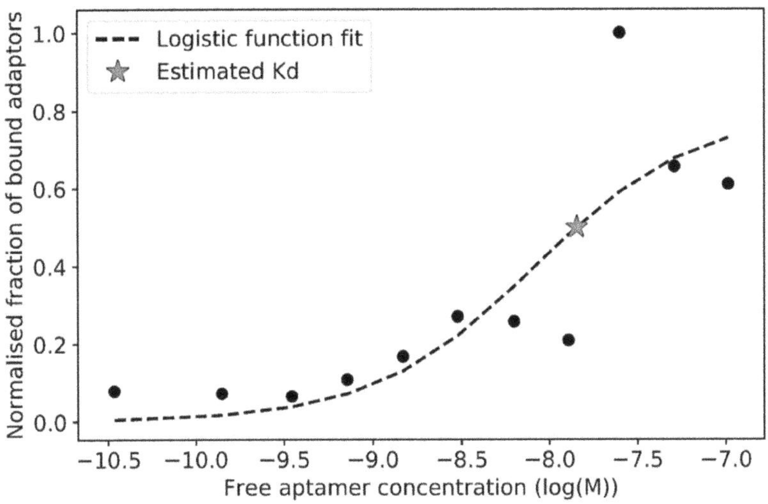

FIGURE 4.13 Binding curve for the estimation of the dissociation constant K_d between aptamers and adaptors. The curve was estimated following the steps described in the Methods section. Reprinted with permission from F. R. Cavallo, K. B. Mirza, S. de Mateo, K. Nikolic, J. Rodriguez-Manzano, and C. Toumazou, "Aptasensor for Quantification of Leptin Through PCR Amplification of Short DNA-Aptamers." *ACS Sensors* 2021 6 (3), 709-715. DOI: 10.1021/acssensors.0c02605.

Results: Validation on human saliva

The assay was validated on human saliva spiked with 1, 10, and 100 ng/mL of leptin standards. A control sample of saliva without added leptin was included to adjust for the influence of leptin already present in the sample. Table 4.4 shows that while there is a three-cycle difference between 100 and 1 ng/mL, recovery is poor due to the late amplification of the control sample (0 ng/mL). Later amplification of the unspiked saliva sample compared to the 0.1 ng/mL standard indicates a suboptimal specificity of the aptamer for leptin, as complexes are displaced by interfering proteins present in the saliva sample.

Discussion

Aptamers have several advantages compared with antibodies for the detection of molecular targets: smaller molecular size, in vitro synthesis, and greater tolerance to temperature. A major advantage of nucleic acid-based aptamers is the possibility of amplification through qPCR, thus resulting in very low limits of detection.

Despite these advantages, commercial aptamer detection kits are still not widely available, and much of the work involving aptamer-based detection has not made it past the research stage. There are multiple reasons behind this. While aptamers have a clear edge over antibodies in terms of storage, preparation, and methodology, comparatively, antibodies tend to have better affinity to the target. This has led to

TABLE 4.4
Recovery and interference measures for leptin in human saliva

Spiked concentration (ng/mL)	C_t mean ± SE	Detected concentration (ng/mL) (mea ± SE)	Recovery
100	47.27	3.63	4%
10	43.98 ± 0.53	5.36 ± 0.39	54%
1	42.76 ± 1.7	0.4 ± 0.3	40%
0.1	30.47	0	0%
0	41 ± 1.69	N/A	N/A

Reprinted with permission from F. R. Cavallo, K. B. Mirza, S. de Mateo, K. Nikolic, J. Rodriguez-Manzano, and C. Toumazou, "Aptasensor for Quantification of Leptin Through PCR Amplification of Short DNA Aptamers." *ACS Sensors* 2021 6 (3), 709-715. DOI:10.1021/acssensors.0c02605.

The Ct and detected concentrations are presented as means and standard errors.

challenges in repeatability when aptamers are used for quantification.[171] There have also been suggestions that the performance of antibodies is still better than aptamers if sensitivity, repeatability, and stability are combined together.[172] However, these considerations apply to lab-based environments. For detection in the field, aptamers seem a better choice than antibodies due to low detection limits, molecular amplification capability, and greater robustness to changes in ambient conditions such as temperature.

This case study presented a method for quantitative aptamer-based sensing using DNA amplification and "on-site" calibration of aptamers. In order to make aptasensors' performance comparable to immunosensors', assay optimisation is key. However, aptamer-based assays are difficult to optimise due to the lack of performance measures applied at each step of the protocol, as the performance is evaluated at the end by measuring the response of aptamers to the target.[173] The ability to quantify short-DNA aptamers—which cannot be amplified directly with PCR due to the low number of bases—can allow the optimisation of the assay at every step, and this need guided the central design choice for the assay described here. In fact, when developing the method with magnetic beads, there was a need to quantify the amount of aptamers that did not bind to the beads for optimisation purposes. However, given the short sequence (26bp), the aptamer could not be amplified directly and consequently, a method to quantify them indirectly was developed. The amplification of the aptamers happens by hybridisation with the adaptor and elongation of the complex into double-stranded DNA, as shown in Figure 4.6. The simplified protocol without magnetic beads presented was subsequently developed with the elongated complexes as the target analyte for leptin quantification.

The aptasensor that was developed provides an improvement in reaction times and ease of preparation, compared to other methods for leptin quantification (see Table 4.5), while also using a very low sample volume.

TABLE 4.5
Comparison of leptin quantification methods

Reference	Assay type	LoD (pg/mL)	K_d (M)	HoT*	TtR (hrs)	Sample (uL)
Imagawa[a]	ELISA	0.78	$83 \cdot 10^{-12}$	N/R	15+	100
He[b]	Chemiluminescent immunosensor	0.3	N/R	N/R	44+	100
Tanaka[c]	Waveguide-mode sensor	10^5	N/R	N/R	24+	400
Dong[d]	Electrochemical immunosensor	30	N/R	N/R	8+	N/R
Cai[e]	Electrochemical immunosensor	0.036	N/R	N/R	78+	N/R
Chen[f]	Electrochemical immunosensor	10^3	N/R	N/R	13+	N/R
Ojeda[g]	Electrochemical immunosensor	0.5	N/R	N/R	1.5+	50
Commercial ELISA	ELISA	15.6	N/R	1.5	3	10, 100***
Our assay	Optical aptasensor	100**	$1.5 \cdot 10^{-6}$	10 min	1.75	10
Our assay w beads	Optical aptasensor	10**	$1.5 \cdot 10^{-6}$	1 hr	3	10

LoD: limit of detection; HoT: hands-on time; TtR: time to result; N/R: not reported. *Hands-on time excludes incubation time and optical reading (the qPCR in the proposed methods takes 1 hour), which are included in the time to result.** The LoD is based on the standard curve. ***10 μL for plasma and serum, 100 μL for supernatant.

[a] Imagawa et al., "Development of a Sensitive Elisa for Human Leptin, Using Monoclonal Antibodies."

[b] He et al., "Detection of Human Leptin in Serum Using Chemiluminescence Immunosensor: Signal Amplification by Hemin/G-Quadruplex Dnazymes and Protein Carriers by Fe3o4/Polydopamine/Au Nanocomposites."

[c] Tanaka et al., "Design and Fabrication of Biosensing Interface for Waveguide-Mode Sensor."

[d] Dong et al., "Amperometric Immunosensor Based on Carbon Nanotubes/Chitosan Film Modified Electrodes for Detection of Human Leptin."

[e] Cai et al., "Porous Graphene-Black Phosphorus Nanocomposite Modified Electrode for Detection of Leptin."

[f] Chen, Lei, and Li, "Regenerable Leptin Immunosensor Based on Protein G Immobilized Au-Pyrrole Propylic Acid-Polypyrrole Nanocomposite."

[f] Ojeda et al., "A Disposable Electrochemical Immunosensor for the Determination of Leptin in Serum and Breast Milk."

[g] Ojeda et al., "A Disposable Electrochemical Immunosensor for the Determination of Leptin in Serum and Breast Milk."

Discussion: Mathematical modelling of the aptasensor

To characterise and understand the performance of the aptasensor, the kinetics of the different reactions are shown below (L is leptin, Ap is aptamer, Ad is adaptor).

The first reaction to consider is between the DNA aptamer and leptin:

$$[L]+[Ap]\overset{K_1}{\rightleftharpoons}[ApL] \tag{4.4}$$

$$K_1=\frac{[ApL]}{[ApL]}=\frac{1}{K_{d,Ap}}, \qquad K_{d,Ap}=1.5\pm 0.25\mu\text{M} \tag{4.5}$$

where $K_{d,Ap}$ and K_1 are the dissociation constant of the aptamer and equilibrium constant of the leptin-aptamer reaction, respectively. The dissociation constant $K_{d,Ap}$ indicates the affinity of the aptamer towards leptin.

The second reaction involves the formation of complexes between the aptamer and adaptor:

$$a[Ap]+b[Ad]\overset{K_1}{\rightleftharpoons}c[ApAd] \tag{4.6}$$

$$K_2=\frac{[ApAd]^c}{[Ap]^a[Ad]^b}\overset{a=b=c=1}{=}\frac{[ApAd]}{[Ap][Ad]}=\frac{1}{K_{d,Ad}}, \qquad K_{d,Ad}=14.3\text{ nM} \tag{4.7}$$

where a, b, c are stoichiometric units of aptamer ($[Ap]$), adaptor ($[Ad]$) and aptamer-adaptor complexes ($[AdAp]$) respectively; and $K_{d,Ad}$ and K_2 are the dissociation constant and the equilibrium constant of the aptamer-adaptor reaction. $K_{d,Ad}$ was estimated by fitting a logistic function to the aptamer-adaptor binding curve (Figure 4.13). When leptin is introduced in a test solution containing $[AdAp]$ complexes, leptin will displace the adaptor from the complex and bind to the aptamer if $K_2 \ll K_1$. The reaction between unknown concentrations of leptin and complexes, where $x<1$ is the unknown stoichiometric proportion of leptin ($[L]$) in relation to the amount of complexes, can therefore be modelled by the equation:

$$[ApAd]+x[L]\overset{K_3}{\rightleftharpoons}x[ApL]+(1-x)[ApAd]+x[Ad] \tag{4.8}$$

Assuming all the leptin has reacted with $[AdAp]$ complexes, we get the equilibrium constant of the whole reaction, which depends on the dissociation constants of aptamer-leptin and aptamer-adaptor:

$$K_3 = \frac{[ApL]^x}{[L]^x}\frac{[Ad]^x}{[ApAd]^x} = \left[\frac{K_{d,Ad}}{K_{d,Ap}}\right]^x \tag{4.9}$$

K_3 which determines the affinity of the complex for leptin, is dependent on the proportion x of leptin and the ratio of the dissociation constants $K_{d,Ad}$ and $K_{d,Ap}$. Hence, the higher the proportion of leptin in the test sample, the higher will be the affinity; similarly, lower $K_{d,Ad}$ will result in greater affinity towards leptin.

Discussion: Limitations and future improvements

The main limitation of this aptasensor is the high dissociation constant of the leptin aptamer. Complexes consisting of an aptamer and a single-strand DNA adaptor have been used before with good results attributable to the high affinity of the aptamer to the target molecule.[174] However, in this chapter, the aptamer used has comparatively low affinity for leptin (the dissociation constant K_d is about 1.50 ± 0.25 μM[175]), which limits the lowest limit of detection of this aptasensor. To improve assay sensitivity, an aptamer with higher affinity for leptin can be selected through *in silico* post-SELEX,[176] by which aptamer-protein interactions can be predicted by using ML algorithms to create a DNA library from mutations of the original aptamer. The proposed method can be easily modified to work with different aptamers, as it would only require the modification of the adaptor sequence and an updated pair of primers.

Higher sensitivity can also be achieved by immobilising aptamers or adaptors on a solid support material and removing excess sequences after each hybridisation step; however, this would increase the assay complexity and the preparation time. Sensitivity and specificity can also be improved by implementing a sandwich assay.[177]

Another factor affecting assay performance is complex stability. The dissociation constant of the complexes is affected by a number of factors, including the length of the hybridisation bases, since a greater length of contact between the aptamer and adaptor would render the complexes more stable. Increased complex stability can improve the specificity of the assay, but it may also compromise the binding of the aptamer with leptin, which only occurs if the complex is displaced. Other factors to consider for an optimal compromise between specificity, sensitivity, and assay complexity are incubation temperature, incubation time, and solution ionic strength.[178]

The assay was tested on saliva, which highlighted the need for further optimisation to detect leptin in real-world samples. The presence of inhibitors in saliva is the main factor affecting detection accuracy and sensitivity, as several proteins are present in high concentrations, while other proteins, including leptin, are present in much smaller amounts.[179] The development of an aptamer with higher specificity for leptin could minimise the impact of inhibitors and improve the assay performance in saliva. There are other issues related to the use of saliva as a diagnostic sample, such as viscosity and the presence of food debris, which can be tackled by further sample processing.

The stability of the aptamer-DNA has not been considered in previous works on aptasensors, even though the potential loss of complexes due to changes in ambient conditions is possible. This causes the release of aptamers and the loss of DNA adaptors, and will undermine the repeatability of results. When an unknown leptin sample is introduced, it may also interact with the free aptamers, which would lead to incorrect readings of the leptin concentrations or the number of molecules. This problem is addressed by the "on-site" calibration for the number of complexes formed and lost due to reasons other than the presence of leptin, which is achieved by including a negative control and standard curve in every assay. This leads to better quantification of leptin and will indicate the "starting point" and maximum limit of quantification.

The simplicity and versatility of aptamers in terms of preparation, storage, and methodology make them ideal candidates for out-of-lab, fast detection of proteins. A significant advantage of aptamers includes their ability to form complexes with oligonucleotides, which can be used to detect molecular targets at low concentrations.

This study presented an aptamer-based detection method. Preliminary results indicate accurate, fast detection of leptin, with a time-to-result shorter than commercially available ELISAs. The lowest limit of detection is enough to detect salivary leptin concentrations,[180] but could be improved if an aptamer with a lower *K*d is used.

This study has been previously published: F. R. Cavallo, K. B. Mirza, S. de Mateo, K. Nikolic, J. Rodriguez-Manzano, and C. Toumazou, "Aptasensor for Quantification of Leptin Through PCR Amplification of Short DNA-Aptamers." *ACS Sensors* 2021 6 (3), 709-715. DOI: 10.1021/acssensors.0c02605.

CASE STUDY: DESIGN OF A POINT-OF-CARE DEVICE FOR THE QUANTIFICATION OF LEPTIN

Previous sections have highlighted the importance of POC tests to realise personalised healthcare and to improve the efficiency of health systems, especially in resource-constrained settings. While countless biosensors with high performance and low limits of detection have been developed, translating them into POC platforms comes with its own set of challenges. Some types of biosensors allow for easier translation into POC devices compared to others, and the choice of bioreceptor can influence the performance of the device in less-than-ideal environmental conditions. Previous sections have discussed that, while antibodies are a common choice of bioreceptor for many biosensors, aptamers present several advantages, especially when considered in the context of POC applications. A key advantage of biosensors utilising DNA/RNA aptamers as bioreceptors is the ability to quantify the target of interest through qPCR. In order to adapt qPCR-based biosensors for POC applications, compact and inexpensive qPCR machines have been developed,[181,182,183,184] but lack the automated sample preparation, which is required for a fully automated platform that is independent of specialised facilities or trained personnel.

This case study presents the design of a POC platform for the quantification of leptin using aptasensors and qPCR. The development of the aptasensors referenced here is discussed thoroughly in the previous case study.

As shown in Figure 4.14, the system is composed of three main modules: a sample preparation module (A), a thermocycler (B) for qPCR, and a sensing and data processing module (C). The system automates the aptamer-based quantification method, the subject of the previous case study.

Methods: Aptamer-adaptor complex methodology

The aptamer-based method for ultrasensitive protein quantification is described in detail in the previous case study and briefly presented here. The aptamer and the adaptor are single-stranded DNA, each one being complementary to the other so that they are loosely bound by complementary base pair bonding to form a complex. To form such complexes, adaptors and aptamers are incubated for 15 min at 37°C. Then, the leptin sample is added to the complexes, which are displaced by the leptin molecules, breaking the aptamer-adaptor bond and hybridising to the aptamers; therefore, the aptamer selectively disassociates from the nucleic acid to selectively bind to leptin in the sample. The solution is then incubated with the enzyme polymerase for 10 min at 60°C, which elongates the leftover complexes, thus turning them into double-stranded DNA ready for qPCR amplification. A graphical description of the protocol is presented in Figure 4.15.

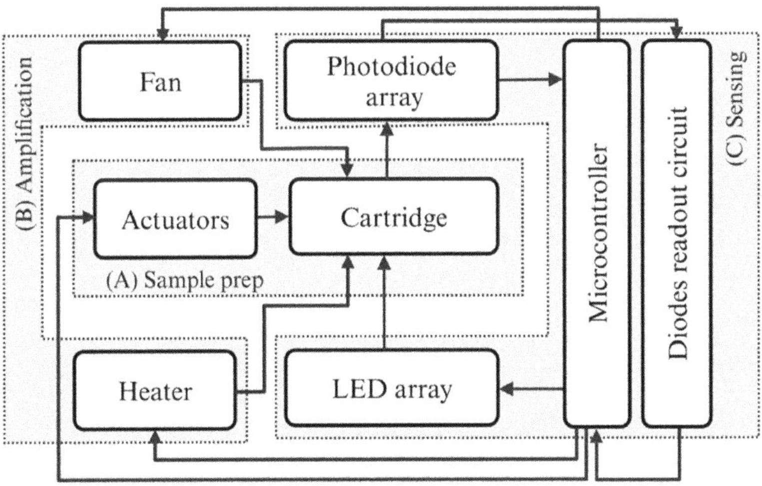

FIGURE 4.14 Block diagram of the protein quantification system. Reprinted with permission from Cavallo F. R., Mirza K. B., de Mateo S., Miglietta L., Rodriguez-Manzano J., Nikolic K., Toumazou C. "A Point-of-Care Device for Fully Automated, Fast and Sensitive Protein Quantification via qPCR." *Biosensors*. 2022; 12(7):537. https://doi.org/10.3390/bios12070537

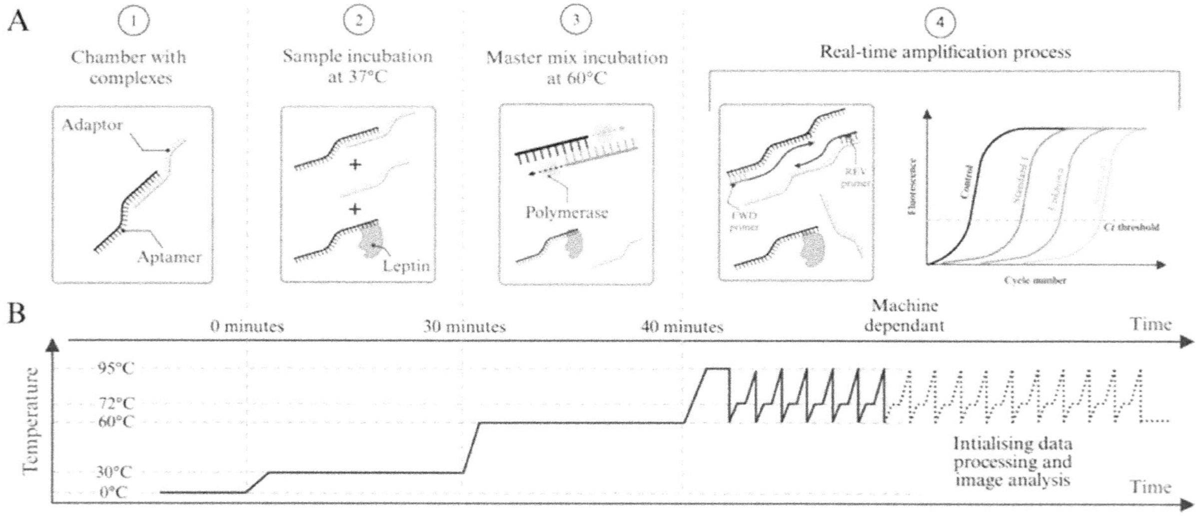

FIGURE 4.15 Methodology for protein sensing: (A1) the chambers are preloaded with aptamer-adaptor complexes; (A2) the sample containing the target protein (leptin) is loaded, and the protein displaces the complexes proportionally to its concentration; (A3) the master mix with the enzyme for complex elongation and PCR is added; (A4, left) the primers are loaded, and the PCR starts; (A4, right) examples of qPCR fluorescence curves. (B) Temperature profile for sample preparation and qPCR, which is pre-set in the machine. The time taken to quantify the sample is machine-dependent as it depends on the amount of aptamer-adaptor complexes formed and retained during the sample preparation steps 1–4. Created in BioRender. Cavallo, F. (2024) BioRender.com/o31t631. Reprinted with permission from Cavallo F. R., Mirza K. B., de Mateo S., Miglietta L., Rodriguez-Manzano J., Nikolic K., Toumazou C. "A Point-of-Care Device for Fully Automated, Fast and Sensitive Protein Quantification via qPCR." *Biosensors*. 2022; 12(7):537. https://doi.org/10.3390/bios12070537

Methods: Sequences and primer designs

The sequences used in the experiments are presented in Table 4.1. The leptin aptamer is from Ashley and Li[185] and the adaptor is designed to have 100bp from the Green Fluorescent Protein (GFP) gene with an extra 26bp for aptamer binding. The primers were designed using Geneious Prime®, and the T_m of the amplification product was determined by the Melting Curve Predictions Software (uMELT) package.[186] Primer specificity was assessed by melting curve analysis after qPCR, which consisted of one cycle at 95°C for 1 min, 40°C for 2 min, followed by a continuous increase in temperature to 90°C at the rate of 0.5°C per second.

Methods: The cartridge

The cartridge shown in Figure 4.16 is an adapted version of the DnaCartridge described by Gibani et al.[187] It is a disposable, sealed, and integrated lab-on-chip device that allows sample-to-result PCR. It consists of two main parts, the amplification unit (AU) and the sample preparation unit, as depicted in Figure 4.16.

The sample preparation unit consists of circumferentially distanced chambers around a rotatable mixing chamber. The chambers contain the aptamer-adaptor complexes, the leptin-binding buffer, the polymerase, and the dried PCR MasterMix (without the primers, which are deposited in dry format into AU's wells). The empty compartments are used to collect waste or for additional buffers, depending on the protocol, which highlights the versatility of the cartridge. The mixing chamber in the middle fits onto a rotating spigot on the NudgeBox (Figure 4.17), with a pneumatic port that helps move reagents between the middle chamber and the rest. The DNA extraction is performed by applying pressure through a syringe. The amplification

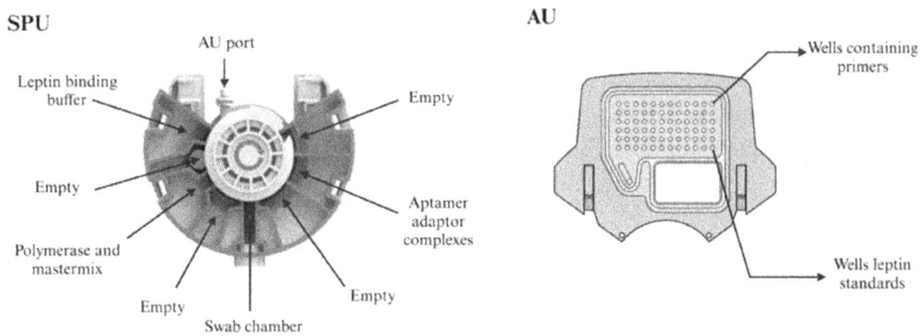

FIGURE 4.16 Sample preparation unit and amplification unit. The sample preparation unit is powered by a motor and transforms the salivary sample into a solution of oligonucleotide that can be quantified through PCR. Once ready, the solution is transferred to the amplification unit, which contains the primers and leptin standards. The high number of wells also allows for multiplexing. Reprinted with permission from Cavallo F. R., Mirza K. B., de Mateo S., Miglietta L., Rodriguez-Manzano J., Nikolic K., Toumazou C. "A Point-of-Care Device for Fully Automated, Fast and Sensitive Protein Quantification via qPCR." *Biosensors.* 2022; 12(7):537. https://doi.org/10.3390/bios12070537

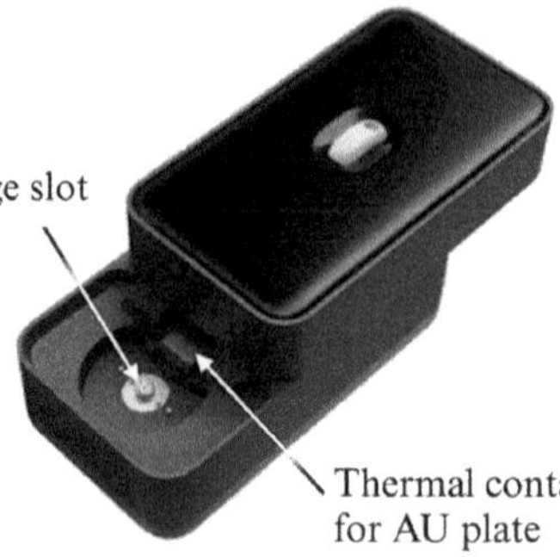

FIGURE 4.17 The NudgeBox. The NudgeBox enables sample processing and quantification of samples with qPCR. The sample processing is done in the cartridge; once inserted in the designated slot, a motor in the NudgeBox completes the preparation step. The processed sample is then transferred to the wells containing the PCR primers; these are placed on the thermal contact that enables the thermal cycling required for PCR. Reprinted with permission from Cavallo F. R., Mirza K. B., de Mateo S., Miglietta L., Rodriguez-Manzano J., Nikolic K., Toumazou C. "A Point-of-Care Device for Fully Automated, Fast and Sensitive Protein Quantification via qPCR." *Biosensors*. 2022; 12(7):537. https://doi.org/10.3390/bios12070537

unit contains the primers that are spotted in nanolitres into the wells and air-dried. Then, to provide replicates and increase accuracy, they are distributed into several wells.

The sample preparation depicted in Figure 4.15 happens in the sample preparation unit, which is preloaded with aptamer-adaptor complexes (Figure 4.15(A1)). After sample loading in the swab chamber, the leptin in the sample displaces the complexes due to high affinity with the aptamers. Subsequently, the solution is mixed with the leptin-binding buffer, which enhances the complex-displacing process. During this step, the solution is incubated at 37°C for 30 minutes (Figure 4.15(A2)). The solution is then mixed and incubated with the PCR MasterMix and the polymerase at 60°C for 10 minutes, which extends the undisplaced complexes to form fully double-stranded sequences (Figure 4.15(A3)). Finally, the solution is transferred to the amplification unit via the amplification unit port, where the solution is mixed with the primers, which are designed to bind specifically to the double-stranded sequences without amplifying free adaptors or aptamers (Figure 4.15(A4)). The AU also contains wells with the leptin standards used for calibration. The AU is placed on a thermal plate (Figure 4.17), which enables PCR amplification.

Methods: The qPCR module

Like other conventional qPCR machines, this module comprises a heating and a cooling element that enables the thermal cycling for the qPCR. However, unlike usual qPCR machines, this module must include a temperature profile for incubation with the sample and with the master mix during the sample preparation (see

Figure 4.15(A3)). After sample preparation, the thermocycler provides the thermal conditions required by the PCR, i.e., pre-incubation at 95°C for 600 seconds and a three-step PCR amplification: DNA denaturation at 95°C for 10 seconds, primer annealing at 60°C for 10 seconds, and DNA elongation at 72°C for 10 seconds. At each annealing step, the blue light of a given wavelength produced by the LEDs ($\lambda \approx$ 497 nm) is absorbed by the SYBR green dye, which emits fluorescence proportional to the target DNA concentration in the chambers. The photodiodes and readout circuit capture the green light ($\lambda \approx$ 520 nm) emitted by the dye.

The design of the qPCR machine has a significant impact on the quantification performance. For example, the speed of the thermal cycling affects the efficiency of the DNA amplification, and noise in the fluorescence detection affects measurement precision.[188]

Methods: Real-time qPCR data processing

The fluorescence data contains the noise introduced during the biochemical process—due to mixing faults and inefficiencies in the biochemical process—named *ProcessFluorescence*. *BackgroundFluorescence* is the noise introduced during the qPCR due to device performance and reaction efficiency, and during the fluorescence reading, due to noise introduced by fluctuations in photodiodes and LED light intensity. The lack of washing steps simplifies the device, which would otherwise require immobilising the aptamers on the wells' surface, a washing buffer, and additional hardware to mix and discard the waste solution. As a result, the free sequences and proteins remain in the chambers and can interfere with the PCR. A negative control is included in the assay and subtracted from the standards and samples to reduce the noise due to the biochemical process (*ProcessFluorescence*). Additionally, the control serves as a validation signal for the assay since it provides a limit on the maximum number of complexes that can be detected. In fact, if the signals produced by the unknown samples show earlier amplification than the negative control, then the free sequences have interfered with the amplification. Consequently, the cartridge is discarded as invalid, and the test is repeated.

PeakFluo (see Algorithm 4.1) is a novel algorithm that processes the fluorescence signal in real time while the PCR is running. *BackgroundFluorescence* represents the background fluorescence produced by the solution in the chamber and is due to device performance. The background fluorescence is detectable when the number of target DNA copies is not enough to produce a visible signal, and therefore *BackgroundFluorescence* is determined as the average fluorescence in the first 10 cycles. From Cycle 11, at each cycle the *BackgroundFluorescence* is subtracted from all signals including the control, and the process noise *ProcessFluorescence* is removed from the standards and unknowns. Given the convex nature of the cleaned fluorescence signal, the process stops once the global minimum is found for each sample, thus reducing the qPCR running time. The instrument performs the qPCR and calculations in real-time, and a standard curve to quantify the samples is generated by the same algorithm run on the standards included in the cartridge.

Algorithm 4.1 PeakFluo: Self-calibrating algorithm for early-cycle protein quantification

Procedure Find chamber noise For $n = 0,\ldots,N$ chambers

For c = 1 to 10:

$$BackgroundFluorescence_n \leftarrow \max\left(f_n\left(c\right)\right)$$

Procedure Find peak fluorescence *F*or $n = 0,\ldots,N$ chambers

$$peak_n = 0$$

While $foundPeak = False$:

For $c = 11$ to $total\ Cycles$:

$$f_n\left(c\right)_{clean} \leftarrow f_n\left(c\right) - BackgroundFluorescence_n - ProcessFluorescence\left(c\right)$$

If $f_n\left(c\right)_{clean} < peak_{n-1}$:

$$peak_n \leftarrow \left|f_n\left(c\right)_{clean}\right|$$

Else:

$$foundPeak = True$$

Procedure Quantify sample

$$m = \frac{peak_{c1} - peak_{c2}}{conc_{c1} - conc_{c2}} \quad c_1 \text{and } c_2 \text{ are standards}$$

$$conc_{sample} = \frac{peak_{sample} - peak_{c_2}}{m} + conc_{c1}$$

Methods: Convolutional neural network for qPCR data classification

An alternative approach to processing the qPCR data is to employ deep learning to analyse the amplification curves as an image. For this purpose, a convolutional neural network (CNN) is trained and validated on 278 qPCR curves generated from the amplification of samples containing leptin standards in various concentrations (0.1, 1, 10 ng/mL). Leptin concentrations are manually labelled into two classes—"high" or "low"—depending on a threshold of 5 ng/mL, which was chosen for its clinical significance. In fact, while leptin-resistant obesity is characterised by increased leptin levels, leptin-deficient obesity is characterised by leptin levels below 5 ng/ml.[189] The dataset composed of 278 qPCR images was collected during the experiments to validate the leptin aptasensor. A total of 201 curves were generated from leptin standards higher than 5 ng/mL and 77 from leptin standards lower than 5 ng/mL.

We trained a six-layer CNN, composed of a convolution layer, max-pooling layer, convolution layer, max-pooling layer, flattening layer, and dense layer (Figure 4.18). The inputs to the CNN are the fluorescence curve images, while the outputs are the labels "high" or "low," classifying the leptin concentrations based on the preset threshold. The model was validated with stratified *k*-fold cross-validation to compensate for the class imbalance. Performance was assessed through classification accuracy and loss.

Methods: Samples for methods development and validation

The biochemical assay was developed and validated with human salivary leptin. PeakFluo and the CNN were trained and validated on leptin standards quantified through the aptamer assay. The qPCR fluorescence signals were obtained with a commercially available PCR machine.

Results

The fluorescence data were collected with a commercially available qPCR machine, using samples prepared manually with the proposed assay.

Figure 4.19 displays the fluorescence signals after the removal of *ProcessFluorescence* and *BackgroundFluorescence*, showing the characteristic peak. The standard curves generated by fitting a linear model to the logarithm of the concentration and the peak fluorescence produced by PeakFluo achieved better R^2 than the standard curves fitted on the C_t values estimated by the commercial

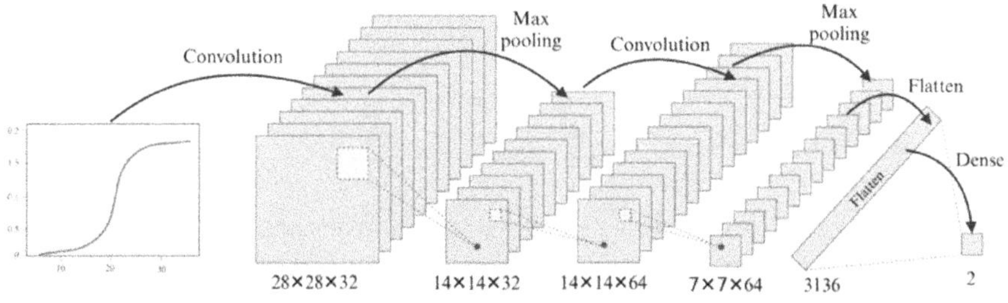

FIGURE 4.18 Architecture of the CNN model. Created in BioRender. Cavallo, F. (2024) BioRender.com/r82a224. Modified with permission from Cavallo F. R., Mirza K. B., de Mateo S., Miglietta L., Rodriguez-Manzano J., Nikolic K., Toumazou C. "A Point-of-Care Device for Fully Automated, Fast and Sensitive Protein Quantification via qPCR." *Biosensors*. 2022; 12(7):537. https://doi.org/10.3390/bios12070537

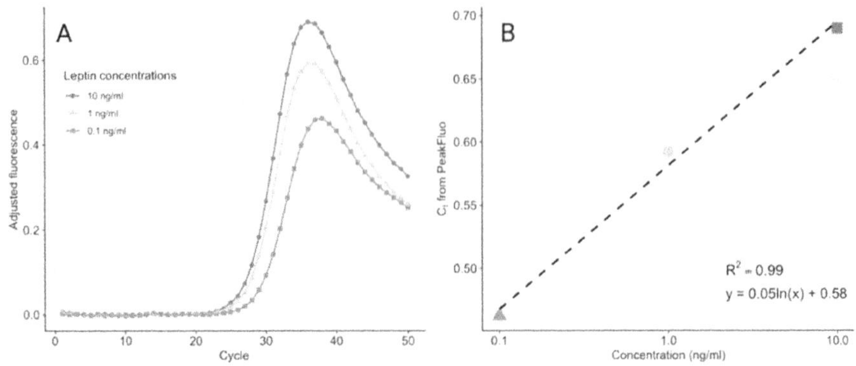

FIGURE 4.19 PeakFluo-generated qPCR fluorescence amplification curves and standard curve. qPCR fluorescence amplification curves after noise removal (A). The noise-adjusted signals decrease at higher cycles as the amplification curves reach a common plateau. The maximum fluorescence at the peak is used to generate the standard curve (B). The data are an example from one experiment using three replicates for each concentration. Reprinted with permission from Cavallo . .R, Mirza K. B., de Mateo S., Miglietta L., Rodriguez-Manzano J., Nikolic K., Toumazou C. "A Point-of-Care Device for Fully Automated, Fast and Sensitive Protein Quantification via qPCR." *Biosensors.* 2022; 12(7):537. https://doi.org/10.3390/bios12070537

TABLE 4.6
Standard curve comparison between PeakFluo and Roche LightCycler 96®

Method	Exp. 1	Exp. 2	Exp. 3
PeakFluo	0.984	0.987	0.77
Roche LightCycler 96®	0.384	0.973	0.53

Reprinted with permission from Cavallo F. R., Mirza K. B., de Mateo S., Miglietta L., Rodriguez-Manzano J., Nikolic K., Toumazou C. "A Point-of-Care Device for Fully Automated, Fast and Sensitive Protein Quantification via qPCR." *Biosensors.* 2022; 12(7):537. https://doi.org/10.3390/bios12070537

software. The results were replicated across several experiments (see Table 4.6 and Figure 4.20). From Figure 4.20, it can be seen that concentrations in the 0.1–10 ng/mL range can be reliably detected, and that the lowest limit of detection is 0.1 ng/mL.

PeakFluo improves linearity by removing the residual fluorescence introduced by the leptin, aptamers and free adaptors remaining in the solution, because they could interfere with the PCR. This residual fluorescence can be removed by including a negative control, thus improving the precision of the measurement. The algorithm

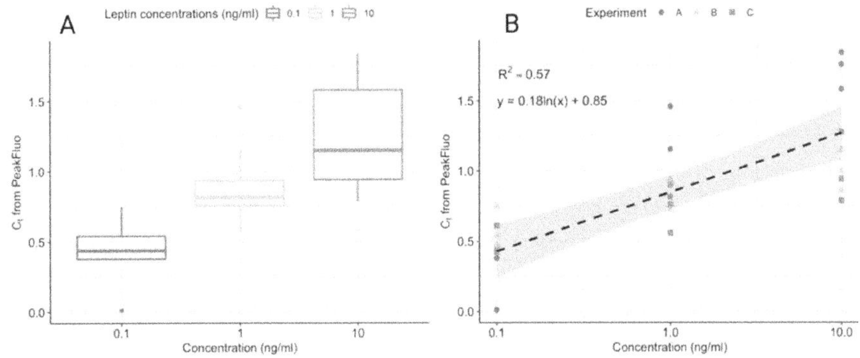

FIGURE 4.20 Representation of variation in leptin concentration across experiments. Boxplot showing the variation in normalised leptin concentrations for three experiments on separate days (A). Linear fit of normalised leptin concentrations for three experiments on separate days with 95% confidence intervals (B). All samples were run in triplicates and data were normalised by dividing each C_t value by the average C_t value. Reprinted with permission from Cavallo F. R., Mirza K. B., de Mateo S., Miglietta L., Rodriguez-Manzano J., Nikolic K., Toumazou C. "A Point-of-Care Device for Fully Automated, Fast and Sensitive Protein Quantification via qPCR." *Biosensors.* 2022; 12(7):537. https://doi.org/10.3390/bios12070537

also reduces the time-to-results, as it allows the PCR to stop around Cycle 35 (Figure 4.19), thus reducing the time-to-result by approximately 20 minutes. Finally, the PeakFluo method operates in real time and it is easily implementable on a portable device.

Results: Convolutional neural network for qPCR data classification

Alongside providing quantitative measurements of protein concentrations, the device can be employed in alternative ways that can provide clinically relevant advice. As discussed in the section "Algorithms for DNA quantification through qPCR," conventional techniques for the analysis of qPCR data involve examining only parts of the fluorescence curve (for example, the exponential phase) to extract relevant features (for example, the C_t value). However, the other phases of the fluorescence curve could hold other information useful to identify target concentrations and differentiate between samples and controls. Therefore, an alternative way of analysing the whole fluorescence curve as an image by utilising convolutional neural networks is proposed. In this way, the complete curve can be analysed, instead of examining only one of its parts. Moreover, with this approach, the calibration steps needed to analyse qPCR data are avoided, such as background removal, calibration, and generation of the standard curve.

Two hundred and seventy-eight qPCR images were collected from a commercially available qPCR machine and manually labelled in two classes based on a threshold of 5 ng/mL (201 high concentration, 77 low concentration). The model was validated

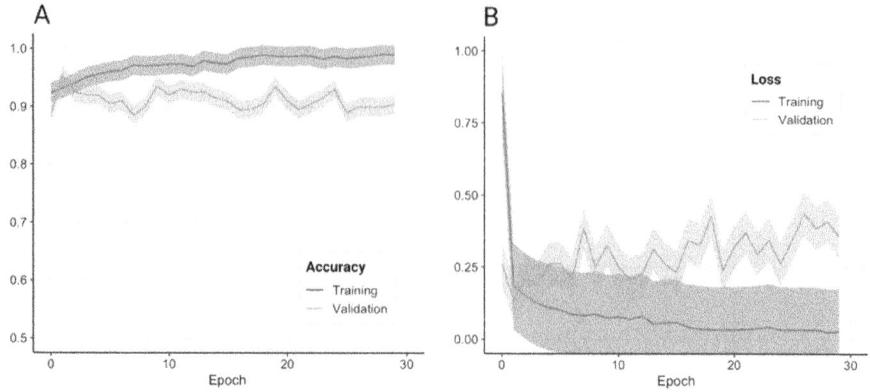

FIGURE 4.21 Model accuracy (A) and loss (B) from ten-fold stratified cross-validation. Data presented as average and standard deviation. Reprinted with permission from Cavallo F. R., Mirza K. B., de Mateo S., Miglietta L., Rodriguez-Manzano J., Nikolic K., Toumazou C. "A Point-of-Care Device for Fully Automated, Fast and Sensitive Protein Quantification via qPCR." *Biosensors*. 2022; 12(7):537. https://doi.org/10.3390/bios12070537

with stratified ten-fold cross-validation to compensate for the class imbalance. The model achieved good training and validation performance (see Figure 4.21).

Discussion

This work presents a fully automated, portable, and rapid system for sensitive protein quantification in a POC setting. The device is based on the novel aptamer-based method for protein quantification described in the previous case study. The aptasensor for the quantification of leptin through qPCR achieves a limit of detection suitable for leptin quantification while using a lower sample volume and significantly less time compared to ELISA, which is the gold standard for protein quantification. Despite the limit of detection being higher than ELISA (100 vs 0.78 pg/mL), the proposed device requires no hands-on time as the sample preparation is fully automated. Additionally, the proposed PeakFluo algorithm can significantly speed up the qPCR process, thereby reducing the time-to-result to under two hours.

As an example of clinical application, this work presents a convolutional neural network trained on qPCR images produced by the proposed device to classify samples based on leptin concentration. The relevance of this approach could be in quickly classifying obese individuals with high or low leptin levels. In fact, while leptin is usually increased in obese patients and is proportional to adipose tissue, leptin is not correlated with BMI for people with the ΔG133 mutation in the *ob* gene, which encodes leptin.[190] Additionally, people with the ΔG133 mutation have significantly lower leptin levels compared to matched controls.[191] Consequently, to administer the right therapy for obese patients, one needs to know either their *ob* genotype or measure their circulating leptin levels. While the former is a viable option, its clinical usefulness is limited as genotypes are immutable. In contrast, monitoring leptin levels can provide ongoing information on the effectiveness of the therapy.

Improvements in neural networks are the object of future work to improve classification performance and validate it on clinical samples. The aim is to provide valuable information to aid clinical decisions and improve outcomes for obese patients. A clinical trial will be needed to clinically validate the device for obesity classification, possibly by employing a three-arm design, including obese subjects with and without the ΔG133 mutation and non-obese subjects. Such a trial would also help assess the clinical usefulness of the device and better characterise its performance in terms of ease of use, speed, and functionality.

This work presents some limitations. Firstly, despite representing a significant improvement compared to currently available ELISA assays, the time-to-result may still be considered long for a rapid POC device. Further development work will aim to shorten the device's detection time. The computational methods, including PeakFluo and the CNN, have been tested on standard leptin concentrations and not on human samples. To validate the methods, tests on different types of human samples will be needed and will be the subject of future work.

This work has been previously published: Cavallo F. R., Mirza K. B., de Mateo S., Miglietta L., Rodriguez-Manzano J., Nikolic K., Toumazou C. "A Point-of-Care Device for Fully Automated, Fast and Sensitive Protein Quantification via qPCR." *Biosensors.* 2022; 12(7):537. https://doi.org/10.3390/bios12070537.

ABBREVIATIONS

AI	Artificial intelligence
CT	Computed tomography
CNN	Convolutional neural network
DNA	Deoxyribonucleic acid
ELISA	Enzyme-linked immunosorbent assay
GFP	Green fluorescent protein
HRP	Horseradish peroxidase
ML	Machine learning
MRI	Magnetic resonance imaging
POC	Point of care
PET	Positron emission tomography
qPCR	Quantitative polymerase chain reaction
SELEX	Systematic evolution of ligands by exponential enrichment
SPR	Surface plasmon resonance
TMB	3,3',5,5' tetramethylbenzidine

NOTES

1. G. S. Firestein, "A Biomarker by Any Other Name," *Nat Rev Rheumatol* 2, no. 12 (2006): 635.
2. P. Dagum, "Digital Biomarkers of Cognitive Function," *NPJ Digit Med* 1, no. 1 (2018): 10.
3. FDA-NIH Biomarker Working Group, "Best (Biomarkers, Endpoints, and other Tools) Resource. 2016," Silver Spring (MD): Food and Drug Administration (US) (2016), https://www.ncbi.nlm.nih.gov/books/NBK326791/ Co-published by National Institutes of Health (US), Bethesda (MD) (Accessed 30/09/2024).

4. National Institute for Health and Care Excellence, "Nice Guideline NG28, Type 2 Diabetes in Adults: Management," National Institute for Health and Care Excellence (2022), https://www.nice.org.uk/guidance/ng28
5. "Nice Guideline Ng136, Hypertension in Adults: Diagnosis and Management," National Institute for Health and Care Excellence (2023), https://www.nice.org.uk/guidance/ng136
6. F. Gundogdu et al., "The Role of Serum CA-125 Levels and CA-125 Tissue Expression Positivity in the Prediction of the Recurrence of Stage III and IV Epithelial Ovarian Tumors (CA-125 Levels and Tissue CA-125 in Ovarian Tumors)," *Arch Gynecol Obstet* 283, no. 6 (2011): 1397–402.
7. F. L. Guest and P. C. Guest, "Point-of-Care Testing and Personalized Medicine for Metabolic Disorders," *Methods Mol Biol* 1735 (2018): 105–14.
8. A. Hernandez-Blanquisett et al., "Current and Emerging Therapies of Her2-Positive Metastatic Breast Cancer," *Breast* 29 (2016): 170–77.
9. T. A. Pearson et al., "Markers of Inflammation and Cardiovascular Disease: Application to Clinical and Public Health Practice: A Statement for Healthcare Professionals from the Centers for Disease Control and Prevention and the American Heart Association," *Circulation* 107, no. 3 (2003): 499–511.
10. M. E. Wasung, L. S. Chawla, and M. Madero, "Biomarkers of Renal Function, Which and When?," *Clin Chim Acta* 438 (2015): 350–57.
11. D. S. Wishart et al., "Markerdb: An Online Database of Molecular Biomarkers," *Nucleic Acids Res* 49, no. D1 (2021): D1259–D67.
12. J. L. Halaas et al., "Weight-Reducing Effects of the Plasma Protein Encoded by the Obese Gene," *Science* 269, no. 5223 (1995): 543–46.
13. J. M. Friedman, "Leptin and the Endocrine Control of Energy Balance," *Nat Metab* 1, no. 8 (2019): 754–64.
14. A. Katsarou et al., "Type 1 Diabetes Mellitus," *Nat Rev Dis Primers* 3, no. 1 (2017): 17016.
15. I. S. Farooqi and S. O'Rahilly, "20 Years of Leptin: Human Disorders of Leptin Action," *J Endocrinol* 223, no. 1 (2014): T63–70.
16. R. A. DeFronzo et al., "Type 2 Diabetes Mellitus," *Nat Rev Dis Primers* 1, no. 1 (2015): 15019.
17. Friedman, "Leptin and the Endocrine Control of Energy Balance."
18. Y. Zhang et al., "Positional Cloning of the Mouse Obese Gene and Its Human Homologue," *Nature* 372, no. 6505 (1994): 425–32.
19. I. S. Farooqi et al., "Effects of Recombinant Leptin Therapy in a Child with Congenital Leptin Deficiency," *N Engl J Med* 341, no. 12 (1999): 879–84.
20. S. Bloom, "Appetite, the Gut and Obesity," *Physiology News Magazine* Summer 2005, no. 59 (2005).
21. N. MacReady, "Leptin: 20 Years Later," *Lancet Diabetes Endocrinol* 2, no. 11 (2014): 864.
22. MacReady, "Leptin: 20 Years Later."
23. H. S. Kassab, "Leptin as a Biomarker in Nutrition and Metabolism," in *Biomarkers in Nutrition*, ed. V. B. Patel and V. R. Preedy, Biomarkers in Disease: Methods, Discoveries and Applications, 1–16 (Cham: Springer International Publishing, 2023).
24. N. Katsiki, D. P. Mikhailidis, and M. Banach, "Leptin, Cardiovascular Diseases and Type 2 Diabetes Mellitus," *Acta Pharmacol Sin* 39, no. 7 (2018): 1176–88.
25. G. V. Childs et al., "The Importance of Leptin to Reproduction," *Endocrinology* 162, no. 2 (2021): bqaa204.
26. K. J. Hong et al., "The Intrafollicular Concentration of Leptin as a Potential Biomarker to Predict Oocyte Maturity in in-Vitro Fertilization," *Sci Rep* 12, no. 1 (2022): 19573.

27. J.-B. Bouillon-Minois et al., "Leptin as a Biomarker of Stress: A Systematic Review and Meta-Analysis," *Nutrients* 13, no. 10 (2021): 3350.
28. T. Ge et al., "Leptin in Depression: A Potential Therapeutic Target," *Cell Death Dis* 9, no. 11 (2018): 1096.
29. T. B. Gonzalez-Castro et al., "Evaluation of Leptin Levels in Serum as a Biomarker for Suicide Behavior: Systematic Review and Meta-Analysis," *Int J Neurosci* 131, no. 1 (2021): 49–55.
30. S. Lee et al., "Plasma Leptin and Alzheimer Protein Pathologies among Older Adults," *JAMA Netw Open* 7, no. 5 (2024): e249539.
31. K. Hamilton et al., "Leptin Prevents Aberrant Targeting of Tau to Hippocampal Synapses Via PI 3 Kinase Driven Inhibition of Gsk3beta," *J Neurochem* 167, no. 4 (2023): 520–37.
32. X. Guo et al., "Revisiting the Effect of Leptin on Alzheimer Disease and Parkinson Disease: The Role of Insulin Sensitivity," *J Clin Endocrinol Metab* 109, no. 9 (2024): 2274–81.
33. Lee et al., "Plasma Leptin and Alzheimer Protein Pathologies among Older Adults."
34. L. G. F. de Carvalho et al., "Circulating Leptin Levels as a Potential Biomarker in Inflammatory Bowel Diseases: A Systematic Review and Meta-Analysis," *Inflamm Bowel Dis* 27, no. 2 (2021): 169–81.
35. M. Nachu, E. C. Dee, and N. Swami, "Equitable Expansion of Preventive Health to Address the Disease and Economic Effect of Ageing Demographics," *Lancet Healthy Longev* 4, no. 4 (2023): e131.
36. A. M. R. Schols et al., "International Definition of a Point-of-Care Test in Family Practice: A Modified E-Delphi Procedure," *Fam Pract* 35, no. 4 (2018): 475–80.
37. S. Shrivastava, T. Q. Trung, and N. E. Lee, "Recent Progress, Challenges, and Prospects of Fully Integrated Mobile and Wearable Point-of-Care Testing Systems for Self-Testing," *Chem Soc Rev* 49, no. 6 (2020): 1812–66.
38. E. Valera et al., "Covid-19 Point-of-Care Diagnostics: Present and Future," *ACS Nano* 15, no. 5 (2021): 7899–906.
39. Ilesh V. Jani, "How Point-of-Care Testing Could Drive Innovation in Global Health," *N Engl J Med* 368, no. 24 (2013): 2319.
40. Meiyun Shang, Jiuchuan Guo, and Jinhong Guo, "Point-of-Care Testing of Infectious Diseases: Recent Advances," *Sens Diagn* 2, no. 5 (2023): 1123–44.
41. Jani, "How Point-of-Care Testing Could Drive Innovation in Global Health."
42. A. Sena-Torralba et al., "Toward Next Generation Lateral Flow Assays: Integration of Nanomaterials," *Chem Rev* 122, no. 18 (2022): 14881–910.
43. C. P. Chan et al., "Development of a Quantitative Lateral-Flow Assay for Rapid Detection of Fatty Acid-Binding Protein," *J Immunol Methods* 279, no. 1–2 (2003): 9–100.
44. J. Vegt, "Development of a User-Friendly App for Assisting Anticoagulation Treatment," *Methods Mol Biol* 1546 (2017): 303–08.
45. D. Du et al., "Integrated Lateral Flow Test Strip with Electrochemical Sensor for Quantification of Phosphorylated Cholinesterase: Biomarker of Exposure to Organophosphorus Agents," *Anal Chem* 84, no. 3 (2012): 1380–85.
46. J. F. Bergua et al., "Lateral Flow Device for Water Fecal Pollution Assessment: From Troubleshooting of Its Microfluidics Using Bioluminescence to Colorimetric Monitoring of Generic Escherichia Coli," *Lab Chip* 21, no. 12 (2021): 2417–26.
47. R. Krska and A. Molinelli, "Rapid Test Strips for Analysis of Mycotoxins in Food and Feed," *Anal Bioanal Chem* 393, no. 1 (2009): 67–71.
48. Jani, "How Point-of-Care Testing Could Drive Innovation in Global Health."
49. R. Zou et al., "Point-of-Care Tissue Analysis Using Miniature Mass Spectrometer," *Anal Chem* 91, no. 1 (2019): 1157–63.

50. S. Banerjee, "Empowering Clinical Diagnostics with Mass Spectrometry," *ACS Omega* 5, no. 5 (2020): 2041–48.
51. S. Arshavsky-Graham and E. Segal, "Lab-on-a-Chip Devices for Point-of-Care Medical Diagnostics," in *Microfluidics in Biotechnology*, ed. Janina Bahnemann and Alexander Grünberger, Advances in Biochemical Engineering/Biotechnology, 247–65 (Springer, 2020).
52. D. Quesada-Gonzalez and A. Merkoci, "Nanomaterial-Based Devices for Point-of-Care Diagnostic Applications," *Chem Soc Rev* 47, no. 13 (2018): 4697–709.
53. N. Gupta et al., "Point-of-Care PCR Assays for Covid-19 Detection," *Biosensors (Basel)* 11, no. 5 (2021): 141.
54. M. M. Gibani et al., "Assessing a Novel, Lab-Free, Point-of-Care Test for SARS-CoV-2 (CovidNudge): A Diagnostic Accuracy Study," *Lancet Microbe* 1, no. 7 (2020): e300–e07.
55. F. Tehrani et al., "An Integrated Wearable Microneedle Array for the Continuous Monitoring of Multiple Biomarkers in Interstitial Fluid," *Nat Biomed Eng* 6, no. 11 (2022): 1214–24.
56. B. Ciui et al., "Wearable Wireless Tyrosinase Bandage and Microneedle Sensors: Toward Melanoma Screening," *Adv Healthc Mater* 7, no. 7 (2018): e1701264.
57. C. Ye et al., "A Wearable Aptamer Nanobiosensor for Non-Invasive Female Hormone Monitoring," *Nat Nanotechnol* 19, no. 3 (2024): 330–37.
58. S. Kadian et al., "Recent Advancements in Machine Learning Enabled Portable and Wearable Biosensors," *Talanta Open* 8 (2023).
59. J. I.. de Oliveira Filho et al., "Empowering Electrochemical Biosensors with Ai: Overcoming Interference for Precise Dopamine Detection in Complex Samples," *Adv Intell Syst* 5, no. 10 (2023): 2300227.
60. K. Zhang et al., "Machine Learning-Reinforced Noninvasive Biosensors for Healthcare," *Adv Healthc Mater* 10, no. 17 (2021): e2100734.
61. A. R. Khan et al., "Point-of-Care Testing: A Critical Analysis of the Market and Future Trends," *Front Lab Chip Technol* 3 (2024): 1394752.
62. D. Lingervelder et al., "Health Economic Evidence of Point-of-Care Testing: A Systematic Review," *Pharmacoecon Open* 5, no. 2 (2021): 157–73.
63. A. El-Osta et al., "Does Use of Point-of-Care Testing Improve Cost-Effectiveness of the NHS Health Check Programme in the Primary Care Setting? A Cost-Minimisation Analysis," *BMJ Open* 7, no. 8 (2017): e015494.
64. P. C. Guest, "Multiplex Biomarker Approaches to Enable Point-of-Care Testing and Personalized Medicine," in *Multiplex Biomarker Techniques: Methods and Applications*, ed. Paul C. Guest, 311–15 (New York: Springer New York, 2017).
65. G. Hayward et al., "Point-of-Care Blood Tests During Home Visits by Out-of-Hours Primary Care Clinicians; a Mixed Methods Evaluation of a Service Improvement," *BMJ Open* 10, no. 1 (2020): e033428.
66. S. Shrivastava, T. Q. Trung, and N.-E. Lee, "Recent Progress, Challenges, and Prospects of Fully Integrated Mobile and Wearable Point-of-Care Testing Systems for Self-Testing," *Chem Soc Rev* 49, no. 6 (2020): 1812–66.
67. J. Wiencek and J. Nichols, "Issues in the Practical Implementation of Poct: Overcoming Challenges," *Expert Rev Mol Diagn* 16, no. 4 (2016): 415–22.
68. Shrivastava, Trung, and Lee, "Recent Progress, Challenges, and Prospects of Fully Integrated Mobile and Wearable Point-of-Care Testing Systems for Self-Testing."
69. J. Howick et al., "Current and Future Use of Point-of-Care Tests in Primary Care: An International Survey in Australia, Belgium, the Netherlands, the UK and the USA," *BMJ Open* 4, no. 8 (2014): e005611.

70. Hayward et al., "Point-of-Care Blood Tests During Home Visits by Out-of-Hours Primary Care Clinicians; a Mixed Methods Evaluation of a Service Improvement."
71. A. Pasquarelli, "Introduction," in *Biosensors and Biochips*, ed. Alberto Pasquarelli, Learning Materials in Biosciences, 1–18 (Cham: Springer International Publishing, 2021).
72. E. R. Kim et al., "Biosensors for Healthcare: Current and Future Perspectives," *Trends Biotechnol* 41, no. 3 (2023): 374–95.
73. Kim et al., "Biosensors for Healthcare: Current and Future Perspectives."
74. V. Naresh and N. Lee, "A Review on Biosensors and Recent Development of Nanostructured Materials-Enabled Biosensors," *Sensors (Basel)* 21, no. 4 (2021): 1109.
75. Naresh and Lee, "A Review on Biosensors and Recent Development of Nanostructured Materials-Enabled Biosensors."
76. B. K. Van Weemen and A. H. Schuurs, "Immunoassay Using Antigen-Enzyme Conjugates," *FEBS Lett* 15, no. 3 (1971): 232–36.
77. S. Campuzano and J. M. Pingarron, "Electrochemical Affinity Biosensors: Pervasive Devices with Exciting Alliances and Horizons Ahead," *ACS Sens* 8, no. 9 (2023): 3276–93.
78. A. Nakatsuma et al., "Detection of HIV-1 P24 at Attomole Level by Ultrasensitive ELISA with Thio-NAD Cycling," *PLoS One* 10, no. 6 (2015): e0131319.
79. M. Morikawa et al., "Subattomole Detection of Adiponectin in Urine by Ultrasensitive ELISA Coupled with Thio-NAD Cycling," *Biophys Physicobiol* 12 (2015): 79–86.
80. P. J. Tighe et al., "Elisa in the Multiplex Era: Potentials and Pitfalls," *Proteomics Clin Appl* 9, no. 3–4 (2015): 406–22.
81. S. Sakamoto et al., "Enzyme-Linked Immunosorbent Assay for the Quantitative/Qualitative Analysis of Plant Secondary Metabolites," *J Nat Med* 72, no. 1 (2018): 32–42.
82. Sakamoto et al., "Enzyme-Linked Immunosorbent Assay for the Quantitative/Qualitative Analysis of Plant Secondary Metabolites."
83. D. Liu et al., "A Fully Integrated Distance Readout Elisa-Chip for Point-of-Care Testing with Sample-in-Answer-out Capability," *Biosens Bioelectron* 96 (2017): 332–38.
84. S. D. de Ferranti et al., "Inflammation and Changes in Metabolic Syndrome Abnormalities in US Adolescents: Findings from the 1988–1994 and 1999–2000 National Health and Nutrition Examination Surveys," *Clin Chem* 52, no. 7 (2006): 1325–30.
85. C. K. Hsu et al., "Paper-Based ELISA for the Detection of Autoimmune Antibodies in Body Fluid—the Case of Bullous Pemphigoid," *Anal Chem* 86, no. 9 (2014): 4605–10.
86. D. Lee, N. Asmare, and A. F. Sarioglu, "Paper-Based Multi-Well Depletion Elisa," *Lab Chip* 23, no. 2 (2023): 251–60.
87. M. S. Verma et al., "Sliding-Strip Microfluidic Device Enables Elisa on Paper," *Biosens Bioelectron* 99 (2018): 77–84.
88. R. C. Murdock et al., "Optimization of a Paper-Based Elisa for a Human Performance Biomarker," *Anal Chem* 85, no. 23 (2013): 11634–42.
89. B. Pang et al., "Development of a Low-Cost Paper-Based Elisa Method for Rapid Escherichia Coli O157:H7 Detection," *Anal Biochem* 542 (2018).
90. C. Coarsey et al., "Development of a Flow-Free Magnetic Actuation Platform for an Automated Microfluidic Elisa," *RSC Adv* 9, no. 15 (2019): 8159–68.
91. T. Wang et al., "Ultrasensitive Microfluidic Solid-Phase Elisa Using an Actuatable Microwell-Patterned Pdms Chip," *Lab Chip* 13, no. 21 (2013): 4190–97.
92. S. Sun et al., "Elisa-Loc: Lab-on-a-Chip for Enzyme-Linked Immunodetection," *Lab Chip* 10, no. 16 (2010): 2093–100.

93. M. J. Uddin, N. H. Bhuiyan, and J. S. Shim, "Fully Integrated Rapid Microfluidic Device Translated from Conventional 96-Well Elisa Kit," *Sci Rep* 11, no. 1 (2021): 1986.
94. A. Parandakh et al., "3D-Printed Capillaric Elisa-on-a-Chip with Aliquoting," *Lab Chip* 23, no. 6 (2023): 1547–60.
95. F. S. Felix and L. Angnes, "Electrochemical Immunosensors – A Powerful Tool for Analytical Applications," *Biosens Bioelectron* 102 (2018): 470–78.
96. Campuzano and Pingarron, "Electrochemical Affinity Biosensors: Pervasive Devices with Exciting Alliances and Horizons Ahead."
97. Kim et al., "Biosensors for Healthcare: Current and Future Perspectives."
98. E. P. Randviir and C. E. Banks, "A Review of Electrochemical Impedance Spectroscopy for Bioanalytical Sensors," *Anal Methods* 14, no. 45 (2022): 4602–24.
99. S. A. Lim and M. U. Ahmed, "Electrochemical Immunosensors and Their Recent Nanomaterial-Based Signal Amplification Strategies: A Review," *RSC Advances* 6, no. 30 (2016): 24995–5014.
100. Lim and Ahmed, "Electrochemical Immunosensors and Their Recent Nanomaterial-Based Signal Amplification Strategies: A Review."
101. I.-H. Cho et al., "Current Technologies of Electrochemical Immunosensors: Perspective on Signal Amplification," *Sensors* 18, no. 1 (2018): 207.
102. J. Zhao et al., "Simultaneous Electrochemical Immunosensing of Alpha-Fetoprotein and Prostate Specific Antigen Using a Glassy Carbon Electrode Modified with Gold Nanoparticle-Coated Silica Nanospheres and Decorated with Azure a or Ferrocenecarboxylic Acid," *Microchimica Acta* 182, no. 15–16 (2015): 2435–42.
103. I. H. Cho, D. H. Kim, and S. Park, "Electrochemical Biosensors: Perspective on Functional Nanomaterials for on-Site Analysis," *Biomater Res* 24, no. 1 (2020): 6.
104. L. Guo et al., "Electrochemical Protein Biosensors for Disease Marker Detection: Progress and Opportunities," *Microsyst Nanoeng* 10, no. 1 (2024): 65.
105. E. K. Wujcik et al., "Antibody Nanosensors: A Detailed Review," *RSC Adv* 4, no. 82 (2014): 43725–745.
106. R. Gao et al., "Wash-Free Magnetic Immunoassay of the PSA Cancer Marker Using Sers and Droplet Microfluidics," *Lab Chip* 16, no. 6 (2016): 1022–29.
107. S. Arshavsky-Graham et al., "Aptasensors Versus Immunosensors-Which Will Prevail?," *Eng Life Sci* 22, no. 3–4 (2022): 319–33.
108. A. E. Cass and Y. Zhang, "Nucleic Acid Aptamers: Ideal Reagents for Point-of-Care Diagnostics?," *Faraday Discuss* 149 (2011): 49–61.
109. A. D. Ellington and J. W. Szostak, "In Vitro Selection of RNA Molecules That Bind Specific Ligands," *Nature* 346, no. 6287 (1990): 818–22.
110. C. Tuerk and L. Gold, "Systematic Evolution of Ligands by Exponential Enrichment: RNA Ligands to Bacteriophage T4 DNA Polymerase," *Science* 249, no. 4968 (1990): 505–10.
111. J. H. Soh, H.-M. Chan, and J. Y. Ying, "Strategies for Developing Sensitive and Specific Nanoparticle-Based Lateral Flow Assays as Point-of-Care Diagnostic Device," *Nano Today* 30 (2020): 100831.
112. Arshavsky-Graham et al., "Aptasensors Versus Immunosensors-Which Will Prevail?"
113. H. Yu et al., "Improving Aptamer Performance: Key Factors and Strategies," *Mikrochim Acta* 190, no. 7 (2023): 255.
114. R. Wang et al., "A Reusable Aptamer-Based Evanescent Wave All-Fiber Biosensor for Highly Sensitive Detection of Ochratoxin A," *Biosens Bioelectron* 66 (2015): 11–8.
115. G. Murtaza, A. S. Rizvi, and F. Qu, "Aptamer-Based Point of Care Testing Schemes," in *Aptamers for Medical Applications: From Diagnosis to Therapeutics* , ed. Y. Dong, 53–76 (Singapore: Springer, 2021).

116. G. Zhang et al., "A Lateral Flow Strip Based Aptasensor for Detection of Ochratoxin a in Corn Samples," *Molecules* 23, no. 2 (2018): 291.
117. J. Liu, D. Mazumdar, and Y. Lu, "A Simple and Sensitive 'Dipstick' Test in Serum Based on Lateral Flow Separation of Aptamer-Linked Nanostructures," *Angew Chem Int Ed Engl* 45, no. 47 (2006): 7955–59.
118. P. Kumar, K. Birader, and P. Suman, "Development of an Impedimetric Aptasensor for Detection of Progesterone in Undiluted Biological Fluids," *ACS Pharmacol Transl Sci* 6, no. 1 (2023): 92–99.
119. O. Mukama et al., "A Highly Sensitive and Specific Lateral Flow Aptasensor for the Detection of Human Osteopontin," *Talanta* 210 (2020): 120624.
120. S. H. Kim et al., "Specific Detection of Avian Influenza H5N2 Whole Virus Particles on Lateral Flow Strips Using a Pair of Sandwich-Type Aptamers," *Biosens Bioelectron* 134 (2019): 123–29.
121. Y. Wang et al., "Label-Free Microfluidic Paper-Based Electrochemical Aptasensor for Ultrasensitive and Simultaneous Multiplexed Detection of Cancer Biomarkers," *Biosens Bioelectron* 136 (2019): 84–90.
122. J. Li et al., "On-Chip, Aptamer-Based Sandwich Assay for Detection of Glycated Hemoglobins Via Magnetic Beads," *Biosens Bioelectron* 79 (2016): 887–93.
123. R. Chinnappan, Q. Ramadan, and M. Zourob, "An Integrated Lab-on-a-Chip Platform for Pre-Concentration and Detection of Colorectal Cancer Exosomes Using Anti-CD63 Aptamer as a Recognition Element," *Biosens Bioelectron* 220 (2023): 114856.
124. F. Costantini et al., "Fluorescent Label-Free Aptasensor Integrated in a Lab-on-Chip System for the Detection of Ochratoxin A in Beer and Wheat," *ACS Appl Bio Mater* 2, no. 12 (2019): 5880–87.
125. Ye et al., "A Wearable Aptamer Nanobiosensor for Non-Invasive Female Hormone Monitoring."
126. B. Wang et al., "Wearable Aptamer-Field-Effect Transistor Sensing System for Noninvasive Cortisol Monitoring," *Sci Adv* 8, no. 1 (2022): eabk0967.
127. K. Mullis et al., "Specific Enzymatic Amplification of DNA in Vitro: The Polymerase Chain Reaction," *Cold Spring Harb Symp Quant Biol* 51 Pt 1 (1986): 263–73.
128. X. Guo et al., "Development of an Ultrasensitive Aptasensor for the Detection of Aflatoxin B1," *Biosens Bioelectron* 56 (2014): 340–44.
129. H. Modh, T. Scheper, and J. G. Walter, "Detection of Ochratoxin A by Aptamer-Assisted Real-Time PCR-Based Assay (Apta-qPCR)," *Eng Life Sci* 17, no. 8 (2017): 923–30.
130. J. Teng et al., "Rolling Circle Amplification Based Amperometric Aptamer/Immuno Hybrid Biosensor for Ultrasensitive Detection of Vibrio Parahaemolyticus," *Microchimica Acta* 184, no. 9 (2017): 3477–85.
131. F. Wu et al., "An Aptamer-Based Ligation-Triggered Rolling Circle Amplification Strategy for ATP Detection and Imaging in Situ," *J Photochem Photobiol A Chem* 355 (2018): 114–19.
132. S. Fredriksson et al., "Protein Detection Using Proximity-Dependent DNA Ligation Assays," *Nat Biotechnol* 20, no. 5 (2002): 473–77.
133. C. A. Heid et al., "Real Time Quantitative PCR," *Genome Res* 6, no. 10 (1996): 986–94.
134. R. G. Rutledge and C. Cote, "Mathematics of Quantitative Kinetic PCR and the Application of Standard Curves," *Nucleic Acids Res* 31, no. 16 (2003): e93.
135. M. Guescini et al., "A New Real-Time PCR Method to Overcome Significant Quantitative Inaccuracy Due to Slight Amplification Inhibition," *BMC Bioinformatics* 9 (2008): 326.
136. M. W. Pfaffl, "Quantification Strategies in Real-Time PCR," in *A-Z of Quantitative PCR*, ed. S.A. Bustin, 89–113 (La Jolla, CA: International University Line, 2004).

137. J. Tellinghuisen and A. N. Spiess, "Comparing Real-Time Quantitative Polymerase Chain Reaction Analysis Methods for Precision, Linearity, and Accuracy of Estimating Amplification Efficiency," *Anal Biochem* 449 (2014): 76–82.
138. S. Zhao and R. D. Fernald, "Comprehensive Algorithm for Quantitative Real-Time Polymerase Chain Reaction," *J Comput Biol* 12, no. 8 (2005): 1047–64.
139. Guescini et al., "A New Real-Time PCR Method to Overcome Significant Quantitative Inaccuracy Due to Slight Amplification Inhibition."
140. Z. Han et al., "Signal Identification of DNA Amplification Curves in Custom-PCR Platforms," *2021 IEEE International Symposium on Circuits and Systems (ISCAS)* (2021): 1–5.
141. A. Moniri et al., "Framework for DNA Quantification and Outlier Detection Using Multidimensional Standard Curves," *Anal Chem* 91, no. 11 (2019): 7426–34.
142. D. J. Alouani et al., "Specificity of SARS-CoV-2 Real-Time Pcr Improved by Deep Learning Analysis," *J Clin Microbiol* 59, no. 6 (2021): e02959.
143. K. Imagawa et al., "Development of a Sensitive Elisa for Human Leptin, Using Monoclonal Antibodies," *Clin Chem* 44, no. 10 (1998): 2165–71.
144. M. Tanaka et al., "Design and Fabrication of Biosensing Interface for Waveguide-Mode Sensor," *Langmuir* 29, no. 42 (2013): 13111–120.
145. Y. He et al., "Detection of Human Leptin in Serum Using Chemiluminescence Immunosensor: Signal Amplification by Hemin/G-Quadruplex Dnazymes and Protein Carriers by Fe_3O_4/Polydopamine/Au Nanocomposites," *Sens Actuators B: Chem* 221 (2015): 792–98.
146. J. Cai et al., "Porous Graphene-Black Phosphorus Nanocomposite Modified Electrode for Detection of Leptin," *Biosens Bioelectron* 137 (2019): 88–95.
147. W. Chen, Y. Lei, and C. M. Li, "Regenerable Leptin Immunosensor Based on Protein G Immobilized Au-Pyrrole Propylic Acid-Polypyrrole Nanocomposite," *Electroanalysis* 22, no. 10 (2010): 1078–83.
148. F. Dong et al., "Amperometric Immunosensor Based on Carbon Nanotubes/Chitosan Film Modified Electrodes for Detection of Human Leptin," *Int J Electrochem Sci* 9 (2014): 6924–35.
149. I. Ojeda et al., "A Disposable Electrochemical Immunosensor for the Determination of Leptin in Serum and Breast Milk," *Analyst* 138, no. 15 (2013): 4284–91.
150. Q. Zhang et al., "Synthesis of Single-Walled Carbon Nanotubes–Chitosan Nanocomposites for the Development of an Electrochemical Biosensor for Serum Leptin Detection," *Mater Lett* 211 (2018): 348–51.
151. X. Liu et al., "Template-Free Synthesis of Mesoporous Ce_3NbOo_7/CeO_2 Hollow Nanospheres for Label-Free Electrochemical Immunosensing of Leptin," *Sens Actuators B: Chem* 341 (2021): 130005.
152. B. Ozcan and M. K. Sezginturk, "A Novel and Disposable GP-Based Impedimetric Biosensor Using Electropolymerization Process with PGA for Highly Sensitive Determination of Leptin: Early Diagnosis of Childhood Obesity," *Talanta* 225 (2021): 121985.
153. T. Islam et al., "Detection of Leptin Using Electrocatalyst Mediated Impedimetric Sensing," *ACS Biomater Sci Eng* 9, no. 5 (2023): 2170–80.
154. He et al., "Detection of Human Leptin in Serum Using Chemiluminescence Immunosensor: Signal Amplification by Hemin/G-Quadruplex Dnazymes and Protein Carriers by Fe_3O_4/Polydopamine/Au Nanocomposites."
155. Dong et al., "Amperometric Immunosensor Based on Carbon Nanotubes/Chitosan Film Modified Electrodes for Detection of Human Leptin."
156. Cai et al., "Porous Graphene-Black Phosphorus Nanocomposite Modified Electrode for Detection of Leptin."

157. Chen, Lei, and Li, "Regenerable Leptin Immunosensor Based on Protein G Immobilized Au-Pyrrole Propylic Acid-Polypyrrole Nanocomposite."
158. Ojeda et al., "A Disposable Electrochemical Immunosensor for the Determination of Leptin in Serum and Breast Milk."
159. J. Ashley and S. F. Li, "Three-Dimensional Selection of Leptin Aptamers Using Capillary Electrophoresis and Implications for Clone Validation," *Anal Biochem* 434, no. 1 (2013): 146–52.
160. F. R. Cavallo et al., "Aptasensor for Quantification of Leptin through PCR Amplification of Short DNA-Aptamers," *ACS Sens* 6, no. 3 (2021): 709–15.
161. C. Erkmen, G. A. Tiğ, and B. Uslu, "First Label-Free Impedimetric Aptasensor Based on Au NPs/TiO_2 NPs for the Determination of Leptin," *Sensors and Actuators B: Chemical* 358 (2022): 131420.
162. C. Erkmen, G. A. Tig, and B. Uslu, "Nanomaterial-Based Sandwich-Type Electrochemical Aptasensor Platform for Sensitive Voltammetric Determination of Leptin," *Mikrochim Acta* 189, no. 10 (2022): 396.
163. Ashley and Li, "Three-Dimensional Selection of Leptin Aptamers Using Capillary Electrophoresis and Implications for Clone Validation."
164. P. Neuzil et al., "Ultra Fast Miniaturized Real-Time PCR: 40 Cycles in Less Than Six Minutes," *Nucleic Acids Res* 34, no. 11 (2006): e77.
165. Ashley and Li, "Three-Dimensional Selection of Leptin Aptamers Using Capillary Electrophoresis and Implications for Clone Validation."
166. Guo et al., "Development of an Ultrasensitive Aptasensor for the Detection of Aflatoxin B1."
167. C. Zong and J. Liu, "The Arsenic-Binding Aptamer Cannot Bind Arsenic: Critical Evaluation of Aptamer Selection and Binding," *Anal Chem* 91, no. 16 (2019): 10887–93.
168. The or threshold cycle, is the cycle number at which the fluorescence generated within a reaction crosses the fluorescence threshold. The fluorescence threshold is set to be significantly above the background fluorescence.
169. J. Xuan et al., "Cloning, Expression and Purification of DNA-Binding Protein Mvo10b from Methanococcus Voltae," *Protein Expr Purif* 64, no. 2 (2009): 162–6.
170. J. W. Jenzano, S. L. Hogan, and R. L. Lundblad, "Factors Influencing Measurement of Human Salivary Lysozyme in Lysoplate and Turbidimetric Assays," *J Clin Microbiol* 24, no. 6 (1986): 963–7.
171. X. Guo et al., "Aptamer-Based Biosensor for Detection of Mycotoxins," *Front Chem* 8 (2020).
172. X. Guo et al., "Aptamer-Based Biosensor for Detection of Mycotoxins."
173. V. C. F. Dos Santos et al., "Real-Time Pcr for Direct Aptamer Quantification on Functionalized Graphene Surfaces," *Sci Rep* 9, no. 1 (2019): 19311.
174. Guo et al., "Development of an Ultrasensitive Aptasensor for the Detection of Aflatoxin B1."
175. Ashley and Li, "Three-Dimensional Selection of Leptin Aptamers Using Capillary Electrophoresis and Implications for Clone Validation."
176. Q. L. Wang et al., "In Silico Post-Selex Screening and Experimental Characterizations for Acquisition of High Affinity DNA Aptamers against Carcinoembryonic Antigen," *RSC Adv* 9, no. 11 (2019): 6328–34.
177. P. Kalra et al., "Simple Methods and Rational Design for Enhancing Aptamer Sensitivity and Specificity," *Front Mol Biosci* 5 (2018): 41.
178. K. Bielec et al., "Kinetics and Equilibrium Constants of Oligonucleotides at Low Concentrations. Hybridization and Melting Study," *Phys Chem Chem Phys* 21, no. 20 (2019): 10798–807.

179. B. J. Baum et al., "Scientific Frontiers: Emerging Technologies for Salivary Diagnostics," *Adv Dent Res* 23, no. 4 (2011): 360–8.
180. M. Groschl et al., "Identification of Leptin in Human Saliva," *J Clin Endocrinol Metab* 86, no. 11 (2001): 5234–39.
181. C. D. Ahrberg et al., "Handheld Real-Time PCR Device," *Lab Chip* 16, no. 3 (2016): 586–92.
182. C. Koo et al., "Development of a Real-Time Microchip PCR System for Portable Plant Disease Diagnosis," *PLoS One* 8, no. 12 (2013): e82704.
183. G. Mulberry et al., "3D Printing and Milling a Real-Time Pcr Device for Infectious Disease Diagnostics," *PLoS One* 12, no. 6 (2017): e0179133.
184. Neuzil et al., "Ultra Fast Miniaturized Real-Time PCR: 40 Cycles in Less Than Six Minutes."
185. Ashley and Li, "Three-Dimensional Selection of Leptin Aptamers Using Capillary Electrophoresis and Implications for Clone Validation."
186. Z. Dwight, R. Palais, and C. T. Wittwer, "Umelt: Prediction of High-Resolution Melting Curves and Dynamic Melting Profiles of PCR Products in a Rich Web Application," *Bioinformatics* 27, no. 7 (2011): 1019–20.
187. Gibani et al., "Assessing a Novel, Lab-Free, Point-of-Care Test for SARS-CoV-2 (CovidNudge): A Diagnostic Accuracy Study."
188. D. S. Lee, "Real-Time PCR Machine System Modeling and a Systematic Approach for the Robust Design of a Real-Time PCR-on-a-Chip System," *Sensors (Basel)* 10, no. 1 (2010): 697–718.
189. I. S. Farooqi et al., "Partial Leptin Deficiency and Human Adiposity," *Nature* 414, no. 6859 (2001): 34–5.
190. Farooqi et al., "Partial Leptin Deficiency and Human Adiposity."
191. Farooqi et al., "Partial Leptin Deficiency and Human Adiposity."

5 Ethical and regulatory issues in personalised healthcare

INTRODUCTION

Personalised medicine aims to use an individual's phenotypes and genotypes (e.g., molecular profiling, medical imaging, lifestyle data) to tailor the right therapeutic strategy for the right person at the right time, to determine the predisposition to disease, or to deliver timely and targeted prevention.[1] Personalised medicine feeds on the availability of data describing the individual, and as the requirement for rich data sources becomes paramount for the realisation of precision and personalised healthcare, there is a concomitant increase in the creation of vast biobanks containing a multitude of health- and lifestyle-related information. The existence of such comprehensive databases containing sensitive information around individuals opens the door to a wide range of ethical and regulatory concerns, such as data privacy and security, health-related discrimination, and equitable access.[2] These topics will be discussed in greater detail in the sections that follow.

The existence of a comprehensive health database does not simply raise concerns related to the existence of the data, but it also poses additional issues that are specific to health data. For instance, while privacy and security are common issues for all data sources, the possibility of misuse of health-related data extends to its interpretation and the evidence behind its meaning. The use of new biomarkers is an ideal example to illustrate how far-reaching these concerns are. As developments in -omics technologies and new computational methods lead to the continuous discovery of new biomarkers, there are increasing concerns around the clinical usefulness of these biomarkers. These concerns are exacerbated by several challenges around the introduction of new biomarkers in clinical practice, the main one arguably being the need for large prospective studies to validate the robustness and efficacy of such biomarkers.[3] As a result, a gap is created by the prospective availability of new biomarkers and the lack of adoption in clinical care. This gap has been filled by direct-to-consumer (DTC) tests, whose availability has expanded significantly in the past two decades,[4] and whose validity is often a source of concern.[5] Although companies do not claim that results can be used for medical decision making,[6] experts still highlight the need for specialised staff to help patients interpret results from genetic tests, both due to the complexity of interpretation, as well as to help minimise the negative reactions that patients often experience after receiving polygenic risk scores

DOI: 10.1201/9781003464983-5

(PRS) from DTC tests.[7] Nonetheless, there is still a risk that healthcare providers might misinterpret genetic information, leading to unnecessary anxiety or inappropriate medical decisions. This is a very concrete risk, given the perceived lack of training around personalised medicine, as reported, for example, by a study with 100 UK gastroenterology trainees, most of whom believed that their training had not prepared them to practise personalised medicine.[8] Further highlighting this perceived lack of training, a systematic review found that only a few physicians consider themselves equipped to interpret genetic test results and many reported the need for more education.[9]

At the same time, the appropriateness of the care provided is dependent on high-quality studies generating evidence that is directly applicable to clinical practice.

The use of biomarkers requires molecular and genetic testing capabilities to be readily available, and their increased use inevitably amplifies the occurrence of false positive results, as well as incidental findings. DTC genetic testing has been found to have alarmingly high rates of false positives (40%), often accompanied by misinterpretation of results by the test providers or third-party services.[10] Analysing -omics data can be complex, and interpretation of the results may not always be straightforward, as the link between biomarkers and disease risk could depend on age, family history, ethnicity, lifestyle, and other environmental factors. Genetic counselling is advocated to ensure the clinical validity and utility of these tests, and confirmatory tests are suggested to avoid potential misdiagnoses or inappropriate treatments.[11] The need for further testing raises concerns around the ethics of devolving resources to retest asymptomatic individuals, especially in resource-constrained systems. Concerns around clinical usefulness and validity apply to tests for single variants as well as for PRS, which incorporate variants that do not necessarily have a causal relationship with the disease trait.[12] Incidental findings, i.e., findings that fall outside the primary purpose of a test,[13] are another ethical concern related to precision and personalised medicine and especially to the creation of research biobanks. Managing incidental findings ethically is a source of debate currently lacking specific guidelines.[14,15]

Overdiagnosis and overtreatment are another concern in personalised medicine. Overdiagnosis occurs when asymptomatic abnormalities are diagnosed as conditions that might not have caused symptoms or harm during a person's lifetime, either because they disappear spontaneously, do not progress, or the person dies of other causes before the symptoms manifest.[16] Overdiagnosis can potentially lead to unnecessary medical interventions, overtreatment, and further over-testing[17] As longitudinal research progresses and understanding of the causal mechanisms of diseases increases, it may become possible to predict which abnormalities will produce clinical symptoms. Until then, the risk of overdiagnosis can be minimised by increasing knowledge and training around this topic,[18] and managing it in clinical practice, for example by only targeting serious conditions that have available treatments.[19]

Finally, when personalised medicine is enabled or delivered through digital technologies, including artificial intelligence (AI), liability may not be easy to attribute.[20] For example, if an AI-enabled clinical support system makes a mistake, is the system, or the clinician that failed to recognise the misdiagnosis, at fault? The development of regulations for AI liability is a field of work in constant evolution, but despite progress in recognising liability issues caused by AI, current regulations still present gaps in the case of opaque "black-box" medical AI system.[21]

INEQUALITY AND DISCRIMINATION

Personalised medicine emphasises the importance of individual differences in achieving targeted prevention and treatment of diseases. However, individuals are not islands, and their health depends not only on individual-specific features such as genetics or behaviour, but also on the socioeconomic, social, and structural contexts, including the healthcare system, in which they live. Issues around the ethical implementation of personalised medicine are centred on the lack of cohort diversity and representativeness of data sources and studies, on the bias that predictive models introduce when trained on biased datasets, and on the social and structural conditions necessary so that everyone benefits from personalised medicine equally. Each aspect is covered in more detail in the sections that follow.

EQUITY AND ACCESS TO PERSONALISED HEALTHCARE

Equity in healthcare is the absence of disparities between groups with different levels of underlying social advantage or disadvantage.[22] The World Health Organization (WHO) defines health equity as "the absence of unfair, avoidable or remediable differences among groups of people, whether those groups are defined socially, economically, demographically, or geographically or by other dimensions of inequality."[23] Consequently, health inequities are differences in health that are unnecessary, avoidable, unfair, and unjust.[24] As health disparities are inequitable if they are systematically associated with social disadvantage, social, economic, and environmental factors—also known as the wider determinants of health[25]—play an important role as determinants of health equity and impact the ability for everyone to benefit from personalised healthcare.

In the UK, the concept of equity was articulated by Aneurin Bevan, the Minister of Health who introduced the National Health Service (NHS) in 1948. The NHS incorporated seven principles to ensure health equity: a service for everyone, free at the point of use, comprehensive, with geographical equality, with the same high standard of care for everyone, with selection on the basis of need and not financial position, and with no incentives for profiting from or exploiting patients.[26] In the context of personalised healthcare, where these principles can be equally applied, equity is undermined when the new paradigm of care, i.e., one in which prevention and treatment are targeted to each patient or patient group, works better for certain people in a way that perpetuates existing inequalities. As an example of how personalised and precision medicine can exacerbate health inequalities, we can consider precision medicine based on genetic research. Genetic associations and PRS are often estimated in populations of European ancestry, meaning that PRS predict risk of disease significantly more accurately in European individuals than non-European.[27] Consequently, the results derived may not be transferable to other populations, opening up issues around the inequitable access of useful genetic risk scores, which may benefit populations that are already on the benefiting end of the health disparity spectrum. To illustrate this, genetic testing for hereditary breast and ovarian cancer, familial hypercholesterolaemia, and Lynch syndrome is significantly lower for marginalised ethnic communities, as well as for disadvantaged groups with lower income, education, and those who are uninsured or underinsured.[28] This happens for a variety of reasons, including lower referral rates, different rates of service uptake, more misdiagnoses or inconclusive results, lack of access to genetic counsellors, inconvenient access to genetic centres, and perceived costs.[29,30] Disparities in the availability of genetic testing are more pronounced in health systems that rely on private insurance, such as in the US, as underserved and minoritised populations are more likely to be uninsured and thus unable to access precision medicine.[31]

With the increasing adoption of digital health solutions, what is commonly known as the "digital divide," a term that describes the gap between those who are digitally enabled and those who are not,[32] has added another layer to the problem of equitable access. Despite digital health holding great promise to reduce inequalities and achieve universal health coverage by, for example, reaching underserved populations, lowering the cost of care, and enabling self-care,[33] paradoxically not everyone may benefit from it. Differences in access to digital health services are seen between age groups, ethnicities, education levels, and socioeconomic status, and are also more pronounced in rural communities and individuals with disabilities or complex needs.[34,35,36] There are several reasons behind lower access to digital health technologies, such as limited access to internet connection and digital devices, poor digital literacy, and distrust in technology.[37,38] Individuals who are white, younger, more educated, and with a higher income have higher digital literacy and engagement with health technology. Since this demographic already experiences better health outcomes, ignoring the digital divide may widen existing inequalities. This phenomenon, whereby interventions benefit already advantaged individuals, is termed "intervention-generated inequality." Even though this term was first used in the context of public health intervention,[39] digital health suffers the same limitation. Mobile health tools for HIV treatment and prevention are an example of how effective digital health solutions would result in benefiting only a subset of the population. Even though the individuals at higher risk of HIV would be helped greatly by digital interventions, the same factors that make them more susceptible to the disease, such as poverty, social exclusion, and stigma, are also associated with lower access to these potentially life-saving technologies.[40] Similar conclusions are found in digital health interventions for a variety of conditions. Digital interventions for increasing physical activity show no evidence of working for individuals of lower socioeconomic status.[41] A study assessing racial disparities in the response to a digital cognitive behavioural intervention for the treatment of insomnia during pregnancy reported no benefits for Black women, despite being effective for non-Hispanic white women.[42] A cross-study evaluating retention in remote digital health studies for Parkinson's, asthma, heart conditions, sleep health, multiple sclerosis, endometriosis, and depression found that minorities and younger populations have the lowest engagement rates, while the group with the highest app use was mainly composed of non-Hispanic whites and older adults.[43] During the Covid-19 pandemic, vulnerable groups—i.e., composed of older adults, migrants, mental health service users, high users of health services, and the unemployed—experienced higher barriers to digital health services access, mainly due to limited digital skills, poor service usability, language limitations, and security and confidentiality concerns.[44]

Digital technologies have become a new determinant of health.[45] Digital determinants of health include concepts such as digital accessibility, digital literacy, user friendliness, interactivity, affordability, availability, algorithmic basis, personalisation, and data poverty.[46]

Digital determinants are strongly related to social determinants of health and affect health outcomes in the same way. In order to avoid worsening inequalities, digital health solutions should strive for equity in access, adoption, adherence, and effectiveness, and studies evaluating their effectiveness should collect and report data on the wider determinants of health of participants.[47] As healthcare is increasingly delivered through digital technologies, the digital divide, if not addressed by developers, researchers, governments, and organisations, will continue to widen health inequalities that are already ingrained in society.[48]

Discrimination

Discrimination refers to the unequal treatment on the basis of personal characteristics, such as sex, age, disability, race, religion or belief, and sexual orientation.[49] Given the amount of data required to power personalised healthcare initiatives, the volume of information available to describe each individual continues to increase, leading to higher risk of misuse and susceptibility to various forms of discrimination. For example, genetic discrimination is defined as "the differential treatment of asymptomatic individuals or their relatives on the basis of their real or assumed genetic characteristics."[50,51] Even though genetic discrimination is not a widespread issue and instances have been recorded mainly in relation to health and life insurance, discrimination on the basis of genetic information is a cause of concern for patients and clinicians.[52] Fear of discrimination is one of the most common reasons to forgo genetic testing,[53] resulting in potential delays in appropriate care.

Other areas of particular concern are those of law enforcement, immigration, and national defence. Several countries, including China, the US, and Thailand, are known to have carried out DNA profiling for law enforcement and security purposes.[54]

While there are codes and regulations addressing these concerns—for instance, the UK's Code on Genetic Testing and Insurance,[55] and the US's Genetic Information Nondiscrimination Act of 2008[56]—they vary substantially across countries and only address discrimination in specific cases.[57] For example, in the US, federal legislation only covers health insurance or employment cases. DTC genetic tests, for example, are not covered by existing regulations, which are not yet equipped to mitigate potential discrimination risks arising from the lack of privacy, confidentiality, and anti-discrimination protections specific to DTC genetic tests.[58] More oversight and pressure from governments and regulators are needed to ensure that the public at large does not oppose genetic data collection, thus undermining research efforts to advance medical knowledge.[59]

Individual behaviour and health inequalities

Throughout this book, personalised healthcare has been discussed with a focus on the prevention and management of chronic diseases, such as type 2 diabetes and cardiovascular disease. Personalised medicine focuses on the individual risk of disease, which, in the case of lifestyle-related conditions, is highly influenced by modifiable lifestyle factors. However, maintaining a good quality of life and health is more difficult for marginalised and disadvantaged groups, and a multitude of factors contribute to health and health-related behaviour. As discussed previously, wider determinants of health span social, economic, and environmental factors that affect which behaviours individuals are more likely to engage in.

Modifiable unhealthy behaviours are not evenly distributed across populations, socioeconomic statuses, and environmental contexts—for example, the availability of fast food, tobacco, and alcohol outlets or sporting facilities is included among the list of environmental and social determinants of disease.[60] This is especially true for type 2 diabetes, where the risk of comorbidities such as cardiovascular disease and all-cause mortality increases with lower socioeconomic status,[61,62] and where drug treatment initiation is delayed in patients with low socioeconomic position.[63] Unhealthy lifestyle and socioeconomic factors are in turn partly a product of structural racism and discrimination[64]—issues that take many years and society at large to fully address. While there are clear benefits to more targeted prevention and treatment strategies, attributing the patient and their personal choices with full responsibility (or blame) for their health without accounting for the wider socioeconomic context risks increasing stigmatisation and diverting resources away from efforts to achieve health equity.[65]

The assumption that poor health is the result of individual choices may exacerbate health inequalities already deeply rooted in societies. In reality, health is the product of a multitude of factors, over which individuals have little or no control. Attributing health outcomes to choice alone, while failing to account for the wider socioeconomic context in which such choices are made, is a dangerous path that can lead to the development of technologies made only for people who can understand and afford them. To avoid developing technologies that widen the health inequality gap, institutions should consider the wider socioeconomic context and develop solutions that are adopted by and benefit all communities. To do so on the public health policy side, Coorey et al.[66] suggest creating a suitable ethical, regulatory, and legislative landscape to ensure data protection and transparency of use. On the research and development side, multistakeholder and multidisciplinary collaboration is critical to ensure the clinical utility and potential acceptability of new technological solutions. Additionally, an important step will be data collection strategies that record information to quantify structural determinants of health disparities, such as lack of access to healthcare, inadequate housing, and environmental factors, and to account for these factors in predictive models used for precision and personalised medicine.[67]

BIASES IN ARTIFICIAL INTELLIGENCE AND MACHINE LEARNING

Bias and discrimination have occupied a large portion of the discourse around the responsible use of AI and machine learning (ML) for healthcare applications. As data quality and representativeness are prerogatives for fair AI systems,[68] the lack of cohort diversity and the bias introduced by AI and ML applications in healthcare are strongly intertwined. The problem of unrepresentative data has serious implications for the health outcomes of poorly represented patient groups. The application of AI in dermatology illustrates this problem well: as datasets to train algorithms for the diagnosis of skin disorders and skin cancer consist mostly of fair-skinned individuals,[69] any algorithm trained on these poorly representative datasets can lead to missed diagnoses for patients with darker skin tones.[70] Genome-wide association studies are another example of unrepresentative data, as they are mostly composed of participants of European ancestry: a review found that over 83% of genome-wide association studies are focused on European populations.[71] As a result, important associations between genes and diseases may be missed in non-European populations,[72] thus affecting the quality of precision medicine for under-represented patient groups and resulting in inequitable access to clinical care informed by genetic research.

Data representativeness goes beyond ensuring that the proportion of people from different ethnicities or different socioeconomic backgrounds included in the training datasets reflects the true population; often, biases are inherent in the data. For example, women are underdiagnosed for heart disease,[73] and South Asian individuals are often underdiagnosed for dementia.[74] This results in intrinsic biases in the data that reinforce existing inequalities. In fact, if an AI model for heart disease risk prediction is trained on a dataset reflecting the common trend of underdiagnosing women, in the absence of bias mitigation measures, men will obtain higher risk scores, thus leading to additional care. These kinds of biases can be prevented by adopting a multidisciplinary approach to model development, by validating the models on minority and low-income groups, and by continuously monitoring solutions for possible unforeseen biases.[75]

It is clear to see that the relationship between equitable access to care and algorithmic bias is a hard cycle to break. Unequitable access leads to biased datasets which, if used to develop personalised healthcare solutions without bias mitigation, lead to even more unequal access to healthcare. Increasing the diversity of the datasets on which medical AI tools are based is a necessary step, but it is not sufficient. To develop bias-free AI-assisted devices, an independent review on equity in medical devices commissioned by the UK Department of Health and Social Care[76] recommended engaging with diverse groups of patients, patient organisations, and the public and involving them in co-design processes. Additionally, the review recommended improving the understanding and knowledge around the issue of equity in AI-assisted medical devices. Finally, to break the structural inequity-algorithmic bias cycle, the authors of the review recommend that researchers, developers, and providers of AI devices be transparent about the diversity, completeness, and accuracy of the data used, and that the tools undergo risk assessment and auditing to identify any existing bias.

PERSUASIVE TECHNOLOGY AND INDIVIDUAL CHOICE

As previously discussed, a person's health is influenced by a multitude of factors, some modifiable, others not. Behaviour is among the modifiable determinants of health and therefore can be changed with behavioural interventions in order to improve health outcomes. Persuasive health technologies are a promising approach to aid the behaviour change process.

Persuasive computers are an interactive technology aimed at changing a person's attitude or behaviour.[77] Persuasive technology (PT) is widely used in a wide range of applications, from education,[78] to health,[79] to the workplace.[80] PT is realised through different platforms, such as social networks, public displays, mobile phones, and games that incorporate psychological, behavioural, and social theories to influence individuals to alter their behaviour.[81] In the field of healthcare, PT is used primarily to enhance digital behaviour change interventions, with examples including interventions for increasing physical activity,[82] reducing sedentary behaviour,[83] managing chronic conditions,[84,85] improving diet,[86,87] and for smoking cessation.[88] While PT used to improve health is usually aligned with a person's best interests, several ethical considerations should be kept in mind when designing interventions that use persuasive strategies.

Firstly, there is the issue of the boundary between persuasion and coercion. Persuasive systems that are truly ethical grant as much autonomy to the individual as possible.[89] However, the user and the provider of the system often don't share the same goals. If the users' goals do not align with the outcome of the PT, or if users do not have the autonomy to discard the goal of the provider; if the PT can block or burden the user's opposition; if users are not aware of the PT influencing their behaviour; then persuasion can shift towards coercion.[90,91] The definition of PT highlights that the behaviour change needs to be voluntary,[92] but as more effective and subtle technologies appear, the difference between persuasion and coercion is not so clear cut. The ultimate way to test if an intervention respects the user's volition is to conduct a person-centred design process, in which the voluntariness and consequently the ethical acceptability of an intervention are constantly informed by the user's perception.[93].

Secondly, PT can influence the individual without requiring their conscious attention. PT can involve well-timed cues or subliminal messages that may light the cognitive load, but it can also eliminate the opportunity for conscious choices. If this happens, the application may not only be of little use due to the lack of personal reflection opportunities, but also unethical. The issue is particularly pronounced when PT is integrated within ambient intelligence and smart environments, whereby individual behaviour is manipulated through subtle cues that bypass their conscious attention.[94,95] Keeping the user informed about the recipient, the commissioner, and the means of delivery, so that users can independently assess the acceptability of an intervention, is the best way to secure consent and respect individual autonomy.[96]

Thirdly, there is the concept of pervasiveness, which can contribute to the perception of PT as a surveillance tool and raise concerns around data privacy and misuse. Pervasiveness is often associated with PT, as digital PT relies on data collected on the user, often in real time. Data can be collected both voluntarily, for example by manually recording one's meals or daily physical activity, or automatically by sensors, social media traffic, internet searches, or GPS. The various aspects that technology can pervade may contribute to its perception as a surveillance tool, especially if paired with a lack of transparency regarding data usage and privacy. As a solution, the user should always be granted autonomy over how data is used and represented.[97]

Finally, special consideration applies when PT is used by vulnerable people, who are the main users of health-related applications of PT. As well as designing PT solutions with stakeholder engagement such that the needs and experiences of vulnerable people are taken into account, PT must respect the person's autonomy.[98] Autonomy should always be upheld in applications using PT, and particularly in vulnerable individuals as autonomy can "counter the sense of powerlessness and loss of agency that is often associated with vulnerability and the risks of objectionable paternalism."[99] Additionally, consent, and the ability to easily rescind it, is key.[100]

PRIVACY AND DATA SECURITY

The protection of health data is an issue of particular importance for society. In the UK, the backlash to the US company Palantir Technologies Inc. winning a contract to deliver the NHS Federated Data Platform in 2023[101] shows how much privacy matters to patients, healthcare professionals, and the public at large. Surveys showed that almost half of adults in England would opt out of data-sharing agreements should the Federated Data Platform be introduced.[102]

Concerns around the misuse of personal data are not unfounded. Over the years, there have been several high-impact examples of inadequate protection of identifiable patient information reported in the media. In 2017, the University of Chicago Medical Center and Google, which partnered to improve predictive analysis in medicine, were sued for failing to strip the data of information that could be used to identify patients.[103] In 2021, an analysis of over 3,700 acute care hospitals in the US found that almost 99% of their websites track and transfer data to third parties, posing a significant risk to confidential patient data.[104] In 2022, two healthcare systems in the US were sued over a data breach due to the use of third-party tracking software scraping patient information.[105] Inadequate protection of privacy is not a prerogative of digital health systems and equally affects consumer products such as mHealth. In a study published in 2021, Tangari et al. assessed over 20,000 mHealth apps and found that very few respect user privacy.[106] Some 88% of apps found in the major app stores can access and share personal data, and 87% made use of third parties to collect user data. Of more concern is the fact that 28% of them did not offer any privacy policy text, and a quarter of user data transmission, which includes geolocation or passwords, violated the terms of the privacy policy. Similarly, other studies reported the lack of respect for user privacy in health and wellness mobile apps,[107,108,109] including among those accredited by the UK's NHS.[110]

Patient data can end up in the hands of unwanted subjects through inadequate consent mechanisms that allow providers to sell patient data to third parties, or through data breaches, enabled by inappropriate security measures that render digital health vulnerable to cyberattacks. Over the past 14 years, there has been an upward trend in healthcare data breaches, with almost 6,000 breaches reported to the US Department of Health and Human Services Office for Civil Rights.[111] Similar to the lack of privacy safeguards, poor security affects consumer products as well as system-wide products. Indeed, studies have reported suboptimal safety measures in mHealth apps, both on the client side[112] and on the server side.[113] Cybersecurity and data breaches are not only a threat to the trust that patients have towards digital health, but can also have catastrophic implications for patient health, as shown by the cyberattack that affected several UK hospitals in mid-2024. The attack resulted in the cancellation of thousands of appointments and created significant blood shortages.[114]

Personalised medicine relies heavily on collecting and analysing individuals' health information. Ensuring the privacy and security of this sensitive information is essential to prevent unauthorised access, data breaches, and potential misuse, and to ensure that patients continue to grant access to their data, without which much of medical research would not be possible.

FUTURE DIRECTIONS AND EMERGING ISSUES

Digital health is a powerful tool in personalised medicine, but, as discussed previously, there are currently important gaps in the efficacy, usefulness, and security of these solutions. Clear and consistent regulations are needed around digital health solutions, especially for those that make use of AI. Consumer apps, in particular, are difficult to regulate as they have low barriers to development, are ubiquitous, can be employed for various use cases, and can directly reach users.[115] While digital health software used for diagnosis, mitigation, treatment, and prevention of disease falls under regulation as a medical device, consumer apps for wellness and fitness are rarely classified as medical devices, although there is a high degree of subjectivity as to what is and is not considered a medical device.[116] As a result, many apps are exempt from regulations, leading to tools that are not rigorously validated[117] and often not secure.[118]

A significant amount of effort is being placed on creating standards for the development and evaluation of digital health apps,[119,120,121,122] and several countries have established frameworks for the distribution of digital health. For example, the UK developed the Digital Technology Assessment Criteria[123] and the NICE Evidence Standards Framework for Digital Health technologies,[124] and Germany's regulatory framework for Digital Health Applications (DiGA),[125] which help clinicians and patients distinguish well-evidenced products from the rest. Additionally, libraries of approved and endorsed apps can help clinicians and patients choose which digital health solutions to use and help spread the adoption of digital health in clinical care.[126]

Digital health is key to enabling personalised healthcare, as it is accessible, convenient, cost-effective, and easily personalisable. However, regulators need to develop consistent frameworks to ensure wide adoption in clinical practice without stifling innovation.[127]

ABBREVIATIONS

AI	Artificial intelligence
DTC	Direct-to-consumer
ML	Machine learning
NHS	National Health Service
PRS	Polygenic risk score
PT	Persuasive technology
WHO	World Health Organization

NOTES

1. Council of the European Union, "Council Conclusions on Personalised Medicine for Patients," *OJEU* C 421 (2015), https://eur-lex.europa.eu/legal-content/EN/TXT/?uri=CELEX:52015XG1217(01).
2. A. Erdmann, C. Rehmann-Sutter, and C. Bozzaro, "Patients' and Professionals' Views Related to Ethical Issues in Precision Medicine: A Mixed Research Synthesis," *BMC Med Ethics* 22, no. 1 (2021): 116.
3. K. Reddy et al., "Subphenotypes in Critical Care: Translation into Clinical Practice," *Lancet Respir Med* 8, no. 6 (2020): 631–43.
4. S. Jiang, L. Liberti, and D. Lebo, "Direct-to-Consumer Genetic Testing: A Comprehensive Review," *Ther Innov Regul Sci* 57, no. 6 (2023): 1190–98.
5. L. Covolo et al., "Internet-Based Direct-to-Consumer Genetic Testing: A Systematic Review," *J Med Internet Res* 17, no. 12 (2015): e279.
6. 23andMe, "Terms of Service," (2022), https://www.23andme.com/en-eu/legal/terms-of-service/ (Accessed 19/08/2024).
7. L. Peck et al., "Why Do People Seek out Polygenic Risk Scores for Complex Disorders, and How Do They Understand and React to Results?" *Eur J Hum Genet* 30, no. 1 (2022): 81–87.
8. I. Al Bakir et al., "Mainstreaming of Genomic Medicine in Gastroenterology, Present and Future: A Nationwide Survey of UK Gastroenterology Trainees," *BMJ Open* 9, no. 10 (2019): e030505.
9. Covolo et al., "Internet-Based Direct-to-Consumer Genetic Testing: A Systematic Review."
10. S. Tandy-Connor et al., "False-Positive Results Released by Direct-to-Consumer Genetic Tests Highlight the Importance of Clinical Confirmation Testing for Appropriate Patient Care," *Genet Med* 20, no. 12 (2018): 1515–21.
11. Tandy-Connor et al., "False-Positive Results Released by Direct-to-Consumer Genetic Tests Highlight the Importance of Clinical Confirmation Testing for Appropriate Patient Care."
12. A. Torkamani, N. E. Wineinger, and E. J. Topol, "The Personal and Clinical Utility of Polygenic Risk Scores," *Nat Rev Genet* 19, no. 9 (2018): 581–90.
13. C. Ells and B. D. Thombs, "The Ethics of How to Manage Incidental Findings," *CMAJ* 186, no. 9 (2014): 655–6.
14. M. Graham et al., "Taking It to the Bank: The Ethical Management of Individual Findings Arising in Secondary Research," *J Med Ethics* 47, no. 10 (2021): 689–96.
15. J. Geiger et al., "Gba/Gbn-Position on the Feedback of Incidental Findings in Biobank-Based Research: Consensus-Based Workflow for Hospital-Based Biobanks," *Eur J Hum Genet* 31, no. 9 (2023): 1066–72.

16. H. Vogt et al., "How Precision Medicine and Screening with Big Data Could Increase Overdiagnosis," *BMJ* 366 (2019): I5270.
17. H. Vogt, S. Green, and J. Broderson, "Precision Medicine in the Clouds," *Nat Biotechnol* 36, no. 8 (2018).
18. M. S. Kale and D. Korenstein, "Overdiagnosis in Primary Care: Framing the Problem and Finding Solutions," *BMJ* 362 (2018): k2820.
19. Vogt et al., "How Precision Medicine and Screening with Big Data Could Increase Overdiagnosis."
20. S. P. Rowland et al., "Digital Health Technology-Specific Risks for Medical Malpractice Liability," *NPJ Digit Med* 5, no. 1 (2022): 87–93.
21. M. N. Duffourc and S. Gerke, "The Proposed EU Directives for AI Liability Leave Worrying Gaps Likely to Impact Medical AI," *NPJ Digit Med* 6, no. 1 (2023): 77.
22. P. Braveman and S. Gruskin, "Defining Equity in Health," *J Epidemiol Community Health* 57, no. 4 (2003): 254–8.
23. World Health Organization, "Health Equity," https://www.who.int/health-topics/health-equity#tab=tab_1 (Accessed 10/09/2024).
24. M. Whitehead, "The Concepts and Principles of Equity and Health," *Int J Health Serv* 22, no. 3 (1992): 429–45.
25. Public Health England, "Health Profile for England: 2018, Research and Analysis," UK Government (2018), Research and analysis.
26. M. Whitehead, "Who Cares About Equity in the NHS?" *BMJ* 308, no. 6939 (1994): 1284–7.
27. A. R. Martin et al., "Clinical Use of Current Polygenic Risk Scores May Exacerbate Health Disparities," *Nat Genet* 51, no. 4 (2019): 584–91.
28. M. J. Khoury et al., "Health Equity in the Implementation of Genomics and Precision Medicine: A Public Health Imperative," *Genet Med* 24, no. 8 (2022): 1630–39.
29. C. Mighton et al., "From the Patient to the Population: Use of Genomics for Population Screening," *Front Genet* 13 (2022): 893832.
30. S. Green, B. Prainsack, and M. Sabatello, "Precision Medicine and the Problem of Structural Injustice," *Med Health Care Philos* 26, no. 3 (2023): 433–50.
31. Green, Prainsack, and Sabatello, "Precision Medicine and the Problem of Structural Injustice."
32. R. Stubbs and M. Brown, "The Digital Divide: Reducing Inequalities for Better Health," (2022), https://publicpolicyprojects.com/wp-content/uploads/2023/07/PPP-Digital-Divide-report-22.pdf.
33. N. O'Brien et al., "Digital Health in Primary Health Care: Current Use and Future Opportunities in the Sub-Saharan African Region," (2023).
34. U. A. Mitchell et al., "The Digital Divide in Health-Related Technology Use: The Significance of Race/Ethnicity," *Gerontologist* 59, no. 1 (2019): 6–14.
35. R. M. Tappen et al., "Digital Health Information Disparities in Older Adults: A Mixed Methods Study," *J Racial Ethn Health Disparities* 9, no. 1 (2022): 82–92.
36. K. E. Woolley et al., "Mapping Inequities in Digital Health Technology within the World Health Organization's European Region Using Progress Plus: Scoping Review," *J Med Internet Res* 25 (2023): e44181.
37. Woolley et al., "Mapping Inequities in Digital Health Technology within the World Health Organization's European Region Using Progress Plus: Scoping Review."
38. T. C. Veinot, H. Mitchell, and J. S. Ancker, "Good Intentions Are Not Enough: How Informatics Interventions Can Worsen Inequality," *J Am Med Inform Assoc* 25, no. 8 (2018): 1080–88.
39. T. Lorenc and K. Oliver, "Adverse Effects of Public Health Interventions: A Conceptual Framework," *J Epidemiol Community Health* 68, no. 3 (2014): 288–90.

40. V. Nittas et al., "Exploring the Equity Impact of Mobile Health-Based Human Immunodeficiency Virus Interventions: A Systematic Review of Reviews and Evidence Synthesis," *Digit Health* 6 (2020): 2055207620942360.
41. M. J. Western et al., "The Effectiveness of Digital Interventions for Increasing Physical Activity in Individuals of Low Socioeconomic Status: A Systematic Review and Meta-Analysis," *Int J Behav Nutr Phys Act* 18, no. 1 (2021): 148.
42. D. A. Kalmbach et al., "Racial Disparities in Treatment Engagement and Outcomes in Digital Cognitive Behavioral Therapy for Insomnia among Pregnant Women," *Sleep Health* 9, no. 1 (2023): 18–25.
43. A. Pratap et al., "Indicators of Retention in Remote Digital Health Studies: A Cross-Study Evaluation of 100,000 Participants," *NPJ Digit Med* 3, no. 1 (2020): 21.
44. A. M. Kaihlanen et al., "Towards Digital Health Equity – A Qualitative Study of the Challenges Experienced by Vulnerable Groups in Using Digital Health Services in the Covid-19 Era," *BMC Health Serv Res* 22, no. 1 (2022): 188.
45. Health The Lancet Digital, "Digital Technologies: A New Determinant of Health," *Lancet Digit Health* 3, no. 11 (2021): e684.
46. S. Chidambaram et al., "An Introduction to Digital Determinants of Health," *PLOS Digit Health* 3, no. 1 (2024): e0000346.
47. Veinot, Mitchell, and Ancker, "Good Intentions Are Not Enough: How Informatics Interventions Can Worsen Inequality."
48. Chidambaram et al., "An Introduction to Digital Determinants of Health."
49. E. Cha, "Inequalities and Multiple Discrimination in Access to and Quality of Healthcare," Luxembourg: European Union Agency for Fundamental Rights (2013), https://fra.europa.eu/en/publication/2013/inequalities-and-multiple-discrimination-access-and-quality-healthcare.
50. P. R. Billings et al., "Discrimination as a Consequence of Genetic Testing," *Am J Hum Genet* 50, no. 3 (1992).
51. M. Otlowski, S. Taylor, and Y. Bombard, "Genetic Discrimination: International Perspectives," *Annu Rev Genomics Hum Genet* 13, no. 1 (2012).
52. Y. Joly and G. Dalpe, "Genetic Discrimination Still Casts a Large Shadow in 2022," *Eur J Hum Genet* 30, no. 12 (2022): 1320–22.
53. T. Yanes et al., "Future Implications of Polygenic Risk Scores for Life Insurance Underwriting," *NPJ Genom Med* 9, no. 1 (2024): 25.
54. F. Forzano et al., "ESHG Warns against Misuses of Genetic Tests and Biobanks for Discrimination Purposes," *Eur J Hum Genet* 29, no. 6 (2021): 894–96.
55. Department of Health and Social Care, "Code on Genetic Testing and Insurance," (2018), https://assets.publishing.service.gov.uk/media/5bd08dd1ed915d78af510220/code-on-genetic-testing-and-insurance.pdf (Accessed 11/09/2024).
56. *Genetic Information Nondiscrimination Act of 2008*, 110–233, 110th United States Congress (21/05/2008).
57. Yanes et al., "Future Implications of Polygenic Risk Scores for Life Insurance Underwriting."
58. J. J. Arias, M. Manchester, and J. Lah, "Direct to Consumer Biomarker Testing for Alzheimer Disease – Are We Ready for the Insurance Consequences?," *JAMA Neurol* 81, no. 2 (2024): 107–08.
59. Forzano et al., "Eshg Warns against Misuses of Genetic Tests and Biobanks for Discrimination Purposes."
60. D. Nunan, D. N. Blane, and M. McCartney, "Exemplary Medical Care or Trojan Horse? An Analysis of the 'Lifestyle Medicine' Movement," *Br J Gen Pract* 71, no. 706 (2021): 229–32.

61. C. A. Jackson et al., "Area-Based Socioeconomic Status, Type 2 Diabetes and Cardiovascular Mortality in Scotland," *Diabetologia* 55, no. 11 (2012): 2938–45.
62. A. Rawshani et al., "Association between Socioeconomic Status and Mortality, Cardiovascular Disease, and Cancer in Patients with Type 2 Diabetes," *JAMA Intern Med* 176, no. 8 (2016): 1146–54.
63. A. C. Falkentoft et al., "Impact of Socioeconomic Position on Initiation of SGLT-2 Inhibitors or GLP-1 Receptor Agonists in Patients with Type 2 Diabetes – A Danish Nationwide Observational Study," *Lancet Reg Health Eur* 14 (2022): 100308.
64. Mary T Bassett and Jasmine D Graves, "Uprooting Institutionalized Racism as Public Health Practice," *American Journal of Public Health* 108, no. 4 (2018): 457.
65. Green, Prainsack, and Sabatello, "Precision Medicine and the Problem of Structural Injustice."
66. G. Coorey et al., "The Health Digital Twin: Advancing Precision Cardiovascular Medicine," *Nat Rev Cardiol* 18, no. 12 (2021): 803–04.
67. Green, Prainsack, and Sabatello, "Precision Medicine and the Problem of Structural Injustice."
68. D. Schwabe et al., "The Metric-Framework for Assessing Data Quality for Trustworthy AI in Medicine: A Systematic Review," *NPJ Digit Med* 7, no. 1 (2024): 203.
69. D. Wen et al., "Characteristics of Publicly Available Skin Cancer Image Datasets: A Systematic Review," *Lancet Digit Health* 4, no. 1 (2022): e64–e74.
70. A. S. Adamson and A. Smith, "Machine Learning and Health Care Disparities in Dermatology," *JAMA Dermatol* 154, no. 11 (2018): 1247–48.
71. M. C. Mills and C. Rahal, "A Scientometric Review of Genome-Wide Association Studies," *Commun Biol* 2, no. 1 (2019): 9.
72. Q. Sun et al., "Analyses of Biomarker Traits in Diverse UK Biobank Participants Identify Associations Missed by European-Centric Analysis Strategies," *J Hum Genet* 67, no. 2 (2022): 87–93.
73. C. Clerc Liaudat et al., "Sex/Gender Bias in the Management of Chest Pain in Ambulatory Care," *Womens Health (Lond)* 14 (2018): 1745506518805641.
74. T. M. Pham et al., "Trends in Dementia Diagnosis Rates in UK Ethnic Groups: Analysis of UK Primary Care Data," *Clin Epidemiol* 10, no. null (2018): 949–60.
75. E. Tat, D. L. Bhatt, and M. G. Rabbat, "Addressing Bias: Artificial Intelligence in Cardiovascular Medicine," *Lancet Digit Health* 2, no. 12 (2020): e635–e36.
76. UK Government, "Equity in Medical Devices: Independent Review," *Department of Health and Social Care* (2024), https://assets.publishing.service.gov.uk/media/65e89e9e62ff48001a87b2d8/equity-in-medical-devices-independent-review-report-web-accessible.pdf#page=73.70 (Accessed 9/12/24).
77. B. J. Fogg, "Persuasive Computers: Perspectives and Research Directions," *Proceedings of the SIGCHI Conference on Human Factors in Computing Systems* (1998): 225–32.
78. S. Devincenzi et al., "Persuasive Technology: Applications in Education" (paper presented at the 2017 IEEE Frontiers in Education Conference (FIE), 18–21 October 2017 2017).
79. R. Orji and K. Moffatt, "Persuasive Technology for Health and Wellness: State-of-the-Art and Emerging Trends," *Health Informatics J* 24, no. 1 (2018): 66–91.
80. K. Wenker, "A Systematic Literature Review on Persuasive Technology at the Workplace," *Patterns* 3, no. 8 (2022): 100545.
81. N. Aldenaini et al., "Persuasive Strategies and Their Implementations in Mobile Interventions for Physical Activity: A Systematic Review," *Int J Hum-Comput Interact* 39, no. 12 (2022): 2292–338.
82. Aldenaini et al., "Persuasive Strategies and Their Implementations in Mobile Interventions for Physical Activity: A Systematic Review."

83. Y. Wang et al., "Persuasive Technology in Reducing Prolonged Sedentary Behaviour at Work: A Systematic Review," *Smart Health* 7–8 (2018): 19–30.
84. M. Berg et al., "Web-Based Intervention for Women with Type 1 Diabetes in Pregnancy and Early Motherhood: Critical Analysis of Adherence to Technological Elements and Study Design," *J Med Internet Res* 20, no. 5 (2018): e160.
85. J. D. Portz et al., "Persuasive Features in Health Information Technology Interventions for Older Adults with Chronic Diseases: A Systematic Review," *Health Technol (Berl)* 6, no. 2 (2016): 89–99.
86. A. Hsu et al., "Persuasive Technology for Overcoming Food Cravings and Improving Snack Choices," in *Proceedings of the SIGCHI Conference on Human Factors in Computing Systems* (Toronto, Ontario, Canada: Association for Computing Machinery, 340312– 2014).
87. J. Pollak et al., "It's Time to Eat! Using Mobile Games to Promote Healthy Eating," *IEEE Pervasive Comput* 9, no. 3 (2010): 21–27.
88. P. O. Rossel et al., "Sas4p: Providing Automatic Smoking Detection for a Persuasive Smoking Cessation Application," *Int J Distrib Sens Netw* 15, no. 11 (2019): 1550147719888167.
89. J. Timmer, L. Kool, and R. van Est, "Ethical Challenges in Emerging Applications of Persuasive Technology," MacTavish, T., Basapur, S. (eds) *Persuasive Technology. PERSUASIVE 2015. Lecture Notes in Computer Science* 9072 (2015): 196–201.
90. Timmer, Kool, and van Est, "Ethical Challenges in Emerging Applications of Persuasive Technology."
91. N. Jacobs, "Two Ethical Concerns About the Use of Persuasive Technology for Vulnerable People," *Bioethics* 34, no. 5 (2020): 519–26.
92. H. Oinas-Kukkonen, "Behavior Change Support Systems: A Research Model and Agenda" (Paper presented at the International Conference on Persuasive Technology, 2010).
93. J. Smids, "The Voluntariness of Persuasive Technology" (Paper Presented at the Persuasive Technology: Design for Health and Safety, 2012).
94. J. Ham, C. Midden, and F. Beute, "Can Ambient Persuasive Technology Persuade Unconsciously?," in *Proceedings of the 4th International Conference on Persuasive Technology*, 1–6 (Claremont, CA: Association for Computing Machinery, 2009).
95. J. Ham and C. Midden, "Ambient Persuasive Technology Needs Little Cognitive Effort: The Differential Effects of Cognitive Load on Lighting Feedback Versus Factual Feedback," *Ploug, T., Hasle, P., Oinas-Kukkonen, H. (eds) Persuasive Technology. PERSUASIVE 2010. Lecture Notes in Computer Science* 6137 (2010): 132–42.
96. R. E. Page and C. Kray, "Ethics and Persuasive Technology: An Exploratory Study in the Context of Healthy Living" (Paper presented at the First International Workshop on Nudge & Influence Through Mobile Devices, Lisbon, 2010).
97. J. Vlieghe and O. De Troyer, "Literature Study on Persuasive Techniques and Technology," *Adaptive Persuasive ICT Tools to Tackle School Burnout Among Youngsters in Brussels, Vrije Universiteit Brussel* (2015), https://wise.vub.ac.be/tickle/wp-content/uploads/2015/12/Report-D3_final.pdf (Accessed 9/12/24).
98. Jacobs, "Two Ethical Concerns About the Use of Persuasive Technology for Vulnerable People."
99. C. Mackenzie, "The Importance of Relational Autonomy and Capabilities for an Ethics of Vulnerability," *Vulnerability: New Essays in Ethics and Feminist Philosophy* (Oxford University Press, 2013), https://doi.org/10.1093/acprof:oso/9780199316649.003.0002.
100. Jacobs, "Two Ethical Concerns About the Use of Persuasive Technology for Vulnerable People."

101. S. Armstrong, "Palantir Gets £480m Contract to Run NHS Data Platform," *BMJ* 383 (2023): 2752.
102. Foxglove and Doctors' Association UK, "The NHS Federated Data Platform and Palantir: 7 Key Risks," (2023).
103. D. Wakabayashi, "Google and the University of Chicago Are Sued over Data Sharing," *The New York Times* (2019), https://www.nytimes.com/2019/06/26/technology/google-university-chicago-data-sharing-lawsuit.html (Accessed 9/12/24).
104. A. B. Friedman et al., "Widespread Third-Party Tracking on Hospital Websites Poses Privacy Risks for Patients and Legal Liability for Hospitals: Study Examines the Widespread Use of Third-Party Tracking on Hospital Websites and the Privacy Risks for Patients and Legal Liability for Hospitals," *Health Aff* 42, no. 4 (2023).
105. D. Muoio and A. Burky, "Advocate Aurora, Wakemed Get Served with Class Action over Meta's Alleged Patient Data Mining," *Fierce Healthcare* (2022), https://www.fiercehealthcare.com/health-tech/report-third-top-hospitals-websites-collecting-patient-data-facebook (Accessed 9/12/24).
106. G. Tangari et al., "Mobile Health and Privacy: Cross Sectional Study," *BMJ* 373 (2021): n1248.
107. Q. Grundy et al., "Data Sharing Practices of Medicines Related Apps and the Mobile Ecosystem: Traffic, Content, and Network Analysis," *BMJ* 364 (2019): l920.
108. K. Huckvale et al., "Unaddressed Privacy Risks in Accredited Health and Wellness Apps: A Cross-Sectional Systematic Assessment," *BMC Med* 13 (2015): 214.
109. A. Papageorgiou et al., "Security and Privacy Analysis of Mobile Health Applications: The Alarming State of Practice," *IEEE Access* 6 (2018): 9390–403.
110. Huckvale et al., "Unaddressed Privacy Risks in Accredited Health and Wellness Apps: A Cross-Sectional Systematic Assessment."
111. Steve Alder, "Healthcare Data Breach Statistics," (2024), https://www.hipaajournal.com/healthcare-data-breach-statistics/ (Accessed 14/09/2024).
112. J. Muthing, T. Jaschke, and C. M. Friedrich, "Client-Focused Security Assessment of Mhealth Apps and Recommended Practices to Prevent or Mitigate Transport Security Issues," *JMIR Mhealth Uhealth* 5, no. 10 (2017): e147.
113. J. Muthing, R. Brungel, and C. M. Friedrich, "Server-Focused Security Assessment of Mobile Health Apps for Popular Mobile Platforms," *J Med Internet Res* 21, no. 1 (2019): e9818.
114. Rebecca Thomas, "NHS Cyber Attack Causing Chaos in London Hits Other Hospitals in UK," *The Independent* (2024), https://www.independent.co.uk/news/health/nhs-russian-cyber-attack-london-hospital-portsmouth-b2559893.html (Accessed 9/12/24).
115. F. Magrabi et al., "Why Is It So Difficult to Govern Mobile Apps in Healthcare?," *BMJ Health Care Inform* 26, no. 1 (2019): e100006.
116. J. Torous, A. D. Stern, and F. T. Bourgeois, "Regulatory Considerations to Keep Pace with Innovation in Digital Health Products," *NPJ Digit Med* 5, no. 1 (2022): 121.
117. C. Prentice et al., "Methods for Evaluating the Efficacy and Effectiveness of Direct-to-Consumer Mobile Health Apps: A Scoping Review," *BMC Digital Health* 2, no. 1 (2024): 31.
118. Tangari et al., "Mobile Health and Privacy: Cross Sectional Study."
119. M. H. Van Velthoven et al., "Digital Health App Development Standards: A Systematic Review Protocol," *BMJ Open* 8, no. 8 (2018): e022969.
120. P. Henson et al., "Deriving a Practical Framework for the Evaluation of Health Apps," *Lancet Digit Health* 1, no. 2 (2019): e52–e54.
121. J. Silberman et al., "Rigorous and Rapid Evidence Assessment in Digital Health with the Evidence Defined Framework," *NPJ Digit Med* 6, no. 1 (2023): 101.

122. M. Kim et al., "The Digital Therapeutics Real World Evidence Framework: An Approach for Guiding Evidence-Based Dtx Design, Development, Testing, and Monitoring," *J Med Internet Res* 26, no. e49208 (2022).
123. NHS England, "Digital Technology Assessment Criteria (DTAC)," https://transform.england.nhs.uk/key-tools-and-info/digital-technology-assessment-criteria-dtac/ (Accessed 15/09/2024).
124. National Institute for Health and Care Excellence, "Evidence Standards Framework for Digital Health Technologies," *NICE Corporate document ECD7* (2018), https://www.nice.org.uk/corporate/ecd7.
125. German Federal Institute for Drugs and Medical Devices, "Diga: Digital Health Applications," https://www.bfarm.de/EN/Medical-devices/Tasks/DiGA-and-DiPA/Digital-Health-Applications/_node.html (Accessed 15/09/2024).
126. W. J. Gordon et al., "Beyond Validation: Getting Health Apps into Clinical Practice," *NPJ Digit Med* 3, no. 1 (2020): 14.
127. Torous, Stern, and Bourgeois, "Regulatory Considerations to Keep Pace with Innovation in Digital Health Products."

Index

D

G

H

I

J

K

L

M

N

O

For Product Safety Concerns and Information please contact our EU representative GPSR@taylorandfrancis.com
Taylor & Francis Verlag GmbH, Kaufingerstraße 24, 80331 München, Germany

www.ingramcontent.com/pod-product-compliance
Ingram Content Group UK Ltd.
Pitfield, Milton Keynes, MK11 3LW, UK
UKHW021721240926
475702UK00029BA/1729
9781032735931